# Basics of Boiler and HRSG Design

**Basics of Boiler and HRSG Design**

**Brad Buecker**

Tulsa, Oklahoma

Copyright 2002 by
PennWell Corporation
1421 S. Sheridan Road
Tulsa, Oklahoma 74112
800-752-9764
sales@pennwell.com
www.pennwell-store.com
www.pennwell.com

Book designed by Clark Bell
Cover photo provided by Black & Veatch Corporation

Library of Congress Cataloging-in-Publication Data
Buecker, Brad
  "Basics of Boiler and HRSG Design/Brad Buecker
    p. cm.
  ISBN 0-87814-795-0
  1. Boilers--Design and construction. I. Title.

  TJ262.5 B84 2002
  621.1'83--dc21

Printed in the United States of America.

1   2   3   4   5   06   05   04   03   02

# DEDICATION

This book is dedicated to the special colleagues with whom it has been a pleasure to work and know for many years. I wish to particularly recognize Todd Hill, Karl Kohlrus, Doug Dorsey, Ellis Loper, Dave Arnold, John Wofford, Ron Axelton, and Sean MacDonald. Not forgotten are all of my other friends at City Water, Light & Power, Burns & McDonnell Engineering, UCB Films and CEDA.

# TABLE OF CONTENTS

# LIST OF FIGURES

# LIST OF TABLES

# LIST OF ACRONYMS

| | |
|---|---|
| ACFB | atmospheric circulating fluidized bed |
| ASME | American Society of Mechanical Engineers |
| ASTM | American Society of Testing & Materials |
| BACT | best available control technology |
| BCC | body-centered cubic |
| BFB | bubbling fluidized bed |
| Btu | British thermal unit |
| CAAA | Clean Air Act Amendment |
| CFB | circulating fluidized bed |
| COHPAC | compact hybrid particulate collector |
| DBA | dibasic acid |
| DNB | departure from nucleate boiling |
| DOE | United States Department of Energy |
| DP | dolomite percentage |
| EPA | Environmental Protection Agency |
| EPRI | Electric Power Research Institute |
| ESP | electrostatic precipitator |
| FAC | flow-assisted corrosion |
| FBHE | fluidized-bed heat exchanger |
| FCC | face-centered cubic |
| FEGT | furnace exit gas temperature |
| FGD | flue gas desulfurization |
| FT | fluid temperature |
| HAP | hazardous air pollutant |
| HCP | hexagonal close packed |
| HHV | higher heating value |
| HP | high pressure |
| HRSG | heat recovery steam generator/generation |
| HT | hemispherical temperature |
| ICGCC | integrated coal gasification combined-cycle |
| IFB | inclined fluidized-bed |
| IP | intermediate pressure |
| IT | initial deformation temperature |
| kV | kilovolt |
| LAER | lowest achievable emission rate |
| LHV | lower heating value |
| LNB | low-NOx burners |
| LP | low pressure |
| MW | megawatt |

| | |
|---|---|
| NAAQS | National Ambient Air Quality Standards |
| NACE | National Association of Corrosion Engineers |
| NiDI | Nickel Development Institute |
| OFA | overfire air |
| PC | pulverized coal |
| PM2.5 | particulate matter less than 2.5 microns in diameter |
| PPM | parts-per-million |
| PPMV | parts-per-million by volume |
| PRB | Powder River Basin |
| RDF | refuse-derived fuel |
| SCR | selective catalytic reduction |
| SD | softening temperature |
| SNCR | selective non-catalytic reduction |
| STP | standard temperature and pressure |
| UNC-CH | University of North Carolina-Chapel Hill |
| WESP | wet electrostatic precipitator |

# FOREWORD

The genesis of this project can be traced to several colleagues who asked me if there was a book on the market describing the basic aspects of fossil-fired steam generator design. I could think of two excellent but very detailed books, Babcock and Wilcox's *Steam* and Combustion Engineering's (now Alstom Power) *Combustion*, but it appeared that a need existed for a condensed version of this material. This book is also "generated" in part by changes in the utility industry, and indeed in other industries—the "do more with less" philosophy. Plants are now being operated by people who have to wear many hats, and may not have extensive training in areas for which they are responsible. The book therefore serves as an introduction to fundamental boiler design for the operator, manager, or engineer to use as a tool to better understand his/her plant. It also serves as a stepping-stone for those interested in investigating the topic even further.

I could not have completed this book without the assistance of many friends who supplied me with important information. These individuals include Mike Rakocy and Steve Stultz of Babcock & Wilcox, Ken Rice and Lauren Buika of Alstom Power, Stacia Howell and Gretchen Jacobson at NACE International, Jim King of Babcock Borsig Power, Jim Kennedy of Foster Wheeler, and Pat Pribble of Nooter Eriksen. All supplied illustrations or granted permission to reproduce illustrations.

The structure of the book is as follows:

- Chapter 1 discusses fundamentals of steam generation and conventional boiler design

- Chapter 2 discusses some of the "newer" (in terms of large-scale use) technologies, including fluidized-bed combustion and heat recovery steam generation (HRSG). For the latter subject, I had the aid of a fine book published by PennWell, *Combined Cycle Gas & Steam Turbine Power Plants, 2nd ed.* For those who really wish to examine combined-cycle operating characteristics in depth, I recommend this book

- Chapter 3 looks at fuel and ash properties

- Chapter 4 examines typical fossil-fuel plant metallurgy. This is very important with regard to plant design and successful operation

- Chapter 5 reviews many important topics regarding air pollution control—a constantly evolving issue. Utility managers will most certainly be faced with new air emissions control challenges in the years and decades to come

I hope you enjoy this book. I spent a number of years at a coal-fired utility, where practical information was of great importance. I have always tried to follow this guideline when writing so the reader can obtain useful data without having to wade through a mountain of extraneous material. I hope this comes through in the book.

Brad Buecker
February 2002
(785) 842-6870
Fax: (785) 842-6944
E-mail: beakertoo@aol.com

# Chapter 1

## Fossil-Fired Boilers– Conventional Designs

## INTRODUCTION

People throughout much of the world have become dependent upon electricity to operate everything from home lighting systems to the most advanced computers. Without electricity, industrial societies would collapse in short order. A very large part of electric power production comes from steam-driven turbine/generators, and even though other sources of energy are becoming more popular, steam-produced electricity will meet our needs for years to come.

Steam also powers many industrial processes that produce goods and services, including foods, pharmaceuticals, steel, plastics, and chemicals. Yet issues related to global warming, acid rain, conservation of resources, and other economic and environmental concerns require that existing plants be operated with the utmost efficiency, while better energy production technologies are being developed.

This chapter provides information about fundamental boiler designs, many of which are still in use today. Knowledge of these basics provides a stepping-stone for understanding newer steam generation technologies, such as the heat recovery steam generator (HRSG) portion of combined-cycle plants.

The steam generating process can be rather complex, especially when electrical generation is part of the network. Consider Figure 1-1. The boiler produces steam to drive both an industrial process and a power-generating turbine. Condensate recovered from the industrial plant is cleaned, blended with condensed steam from the turbine, and the combined stream flows through a series of feedwater heaters and a deaerator to the boiler. The superheater increases steam heat content, which in turn improves turbine efficiency. The turbine itself is an intricate and finely tuned machine, delicately crafted and balanced to operate properly (see Appendix 1-1).

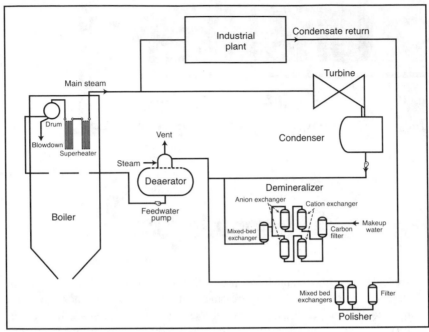

**Fig. 1-1:** Possible water/steam network at a co-generation plant

How did steam-generating units evolve into such complex systems? The process has taken several hundred years.

# Early history of steam generation

The Industrial Revolution of the eighteenth and nineteenth centuries drove a spectacular increase in energy requirements throughout Western Europe, the U.S., and other areas of the world. Some of the industries that blossomed, such as steel making, utilized a great deal of direct heating. However, many processes also required what might be termed *indirect* or *step-wise* heating, in which combustion of fossil fuels in a pressure vessel converts water to steam. It is then transported to the process for energy transfer. Water is used as the energy transfer medium for many logical reasons. It is a very stable substance, available in great abundance, and because of its abundance, is inexpensive.

The first chapter of Babcock & Wilcox's book, *Steam,* outlines the early history of steam generation. The French and British developed practical steam applications in the late 1600s and early 1700s, using steam for food processing and operating water pumps, respectively. The boilers of that time were very simple devices, consisting of kettles heated by wood or charcoal fires.

Technology slowly progressed throughout the 1700s, and by the end of the century and into the early 1800s, several inventors had moved beyond the very basic, and very inefficient, kettle design, developing simple forms of water tube boilers (Fig. 1-2). This period also witnessed the development of fire tube boilers, in which combustion gases flow through boiler tubes with the liquid contained by the storage vessel. The fire tube design had one major disadvantage—the boiler vessel could not handle very high pressures. Many lives were lost due to fire tube boiler explosions in the 1800s, and the design lost favor to water tube boilers. Since the latter dominate the power generation and most of the industrial market, this book will focus exclusively on water tube boilers.

The world changed forever with the invention of practical electrical systems in the early 1880s and development of steam turbines for power generation around the turn of the twentieth century. Ever since, inventors and researchers have worked to improve generation technology in the quest for more efficient electricity production. This chapter looks at conventional boiler types from the late 1950s onward.

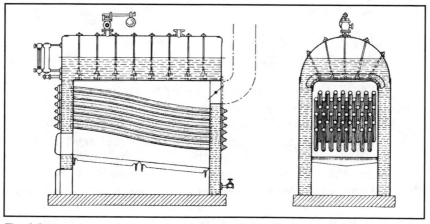

**Fig. 1-2:** An early steam boiler developed by Stephen Wilcox (Reproduced with permission from *Steam*, 40th ed., published by Babcock & Wilcox, a McDermott Company)

# Steam generating fundamentals

This section first examines the basics of heat transfer, beginning with the three major types of energy transfer in nature, providing a basis for understanding heat transfer in a boiler.

# Radiant energy, conduction, and convection

Consider a summer day after sunrise. *Radiant energy* from the sun directly warms the soil. The soil re-radiates some of this energy in the infrared region, but it also heats air molecules that vibrate and agitate other air molecules. This heating process is *conduction*. As the air warms, it rises, and cooler air flows in to take its place. This flow of fluids due to density difference is *convection*.

All three energy transfer mechanisms—radiant energy, conduction, and convection—are at work in a boiler. Radiant heat is obvious—burning fuel emits energy in the form of light and heat waves that travel directly to boiler tubes and transfer energy. Conduction is another primary process wherein the heat produced by the burning fuel greatly agitates air and the combustion-product molecules, which transfer heat to their surroundings. Conduction is also the mode of heat transfer through the boiler tubes; but in this case, the vibrating molecules are those of the tubes. Convection occurs both naturally and mechanically on the combustion and waterside of the boiler. Fans assist convection on the gas side, while waterside convection occurs both naturally and assisted by pumps. Waterside and combustion-side flow circulation are examined in more detail in this chapter and chapter 2.

# Properties of water and steam

In addition to the reasons mentioned earlier for the selection of water as a heating medium, another is its excellent heat capacity. At standard temperature and pressure (STP) of 25°C (77°F) and one atmosphere (14.7 psi), heat capacity is 1 British thermal unit (Btu) per lbm-°F (4.177 kJ per kg-°C). Other physical aspects are also important. Between the freezing and boiling points, any heat added or taken away directly changes the temperature of the liquid. However, at the freezing and boiling points, additional mechanisms come into play. Consider the scenario in which water is heated at normal atmospheric pressure, and the temperature reaches 212°F (100°C). At this point, further energy input does not raise the temperature, but rather is used in converting the liquid to a gas. This is known as the latent heat of vaporization. Thus, it is possible to have a water/steam mixture with both the liquid and vapor at the same temperature. At atmospheric pressure, it takes about 970 Btu to convert a pound of water to steam (2,257 kJ/kg). Once all of the water transforms to steam, additional heating again results in a direct temperature increase. Likewise, when water is cooled to 32°F, additional cooling first converts the water to ice before the temperature drops any lower. This is known as the latent heat of fusion.

As a closed pressure vessel, a boiler allows water to be heated to temperatures much higher than those at atmospheric conditions. For example, in a boiler that operates at 2,400 psig (16.54 mPa), conversion of water to steam occurs at a temperature of 663°F (351°C). Thus, it is possible to add much more heat to the water than at atmospheric pressure. This in turn gives the fluid more potential for work in a heat transfer device. The following discussion of boiler designs illustrates how the boiler components extract energy from burning fuel to produce steam.

## Fundamental boiler design

Figure 1-3 is the simple outline of a natural circulation, drum-type boiler. While this is an elementary diagram, the essentials of water/steam flow are illustrated in this drawing.

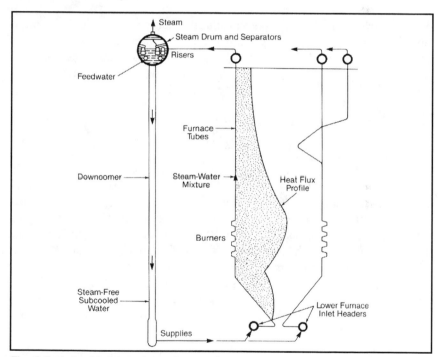

**Fig. 1-3:** A simplified view of water flow in a drum-type, natural-circulation boiler (Reproduced with permission from *Steam*, 40th ed., published by Babcock & Wilcox, a McDermott Company)

Steam generation begins in the waterwall tubes located within the furnace area of the boiler. As the boiler water flowing into the tubes absorbs heat, fluid density decreases and the liquid rises by convection. Conversion to steam begins as the fluid flows upwards through the waterwall tubes, known as risers. (As Appendix 1-2 outlines, a smooth transition of water to steam in the tubes is important.) The water/steam mixture enters the drum, where physical separation occurs with the steam exiting through the main steam line at the top of the drum. The remaining boiler water, plus condensate/feedwater returned to the drum from the turbine or other processes, flows through unheated downcomers to lower waterwall headers.

Many boilers are of the natural circulation type, in which density change is the driving force for movement of water through the boiler. Resistance to flow is primarily due to vertical head friction in the waterwall tubes. The maximum practical pressure for natural circulation units is 2,800 psia (19.31 MPa). At this pressure, the density of water has decreased to about three times that of steam, reducing the boiler's natural circulation capabilities. Contrast this with a 1,200 psia (8.27 Mpa) boiler, where the density ratio of water to steam is 16:1.

A popular design for high-pressure, but still "subcritical" (<3,208 psig) drum units is the forced-circulation type, in which pumps within the downcomer lines mechanically circulate water through the boiler. Additional details regarding these units appear later in this chapter, as well as information on once-through steam generators.

Drum-type boilers have been and still are very popular because the physical separation of the boiler water and steam allows for steady operation and helps prevent steam contamination. Figure 1-4 illustrates a common arrangement of internal drum components. Note the cyclone and secondary steam separators. The cyclones impart a circular motion to the rising steam, which throws entrained moisture to the outside of the canisters where it drains back to the drum. The secondary separators have a chevron vane configuration, and remove water by forcing the steam to make directional changes. Residual water droplets impinge on the vanes and drain back to the drum. A number of different steam separator designs exist, but all serve the same purpose—to remove entrained moisture and prevent carryover of boiler water solids to the superheater and turbine. Damage to separators, poor drum level control, improper water treatment programs, or severe boiler water contamination will allow impurities to enter the steam with potentially dire consequences.

Water entering the drum from the risers may be agitated due to the steaming process in the tubes. Severe turbulence can cause false drum level readings. Figure 1-4 illustrates drum baffle plates (called manifold baffle plates on the figure), which dampen the agitation of the entering water/steam mixture.

Condensate and makeup feed to the boiler enter through the feedwater line. This is a perforated pipe that traverses part of the drum length to ensure a uniform distribution of flow. Feedwater to a utility boiler consists of spent steam recovered

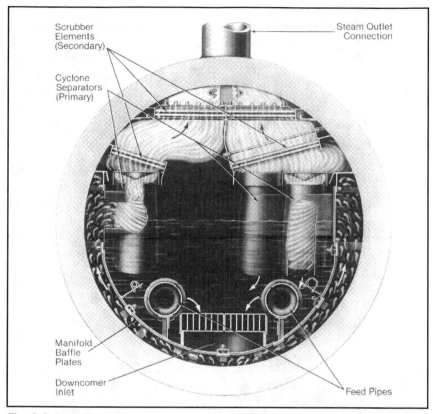

**Fig. 1-4:** Steam drum with steam separators and other internal components (Reproduced with permission from *Steam*, 40th ed., published by Babcock & Wilcox, a McDermott Company)

from the turbine plus a small amount of makeup. Makeup in the range of 1% to 2% is common, and higher percentages suggest major steam leaks. Feedwater in an industrial or co-generation system may come from several different sources, including industrial process returns. Not infrequently, some of the condensate is consumed by the industrial process and must be replenished with fresh makeup. The condensate may be of too poor a quality to be directly introduced to the boiler, and must be cleaned up or dumped. As condensate return percentages decrease, makeup water rates increase.

The chemical feed line to the drum is typically only an inch or so in diameter, as common chemical feed rates are usually slight and are measured in gallons per hour (liters per hour). This line also traverses a portion of the drum to ensure adequate distribution of treatment chemicals.

Continuous water evaporation causes impurities to build up in the boiler water. Potential contaminant sources include the condenser, makeup water system, condensate return lines, and even chemical feed tanks if they are not properly protected from the plant environment. Without some method of boiler bleed-off, impurities will eventually accumulate to a level that causes water and steam chemistry problems. Removal of contaminants is a function of the drum blowdown, which is a small diameter (1" or so) extraction line that resides below the drum water level. While manual blowdown is employed at some plants, automatic blowdown is common—a control system opens a valve when the specific conductivity of the boiler water reaches a preset limit.

A very common drum-boiler design is the two-drum arrangement. An example of a small industrial two-drum boiler is shown in Figure 1-5. The lower drum is referred to as the mud drum. Its primary purpose is to serve as a collection point for solids produced by precipitating chemical treatment programs (see Appendix 1-2). The mud drum usually has a manually operated blowdown. A short bleed-off every day or on some periodic schedule drains precipitates generated by chemical treatments. The operator must be careful not to leave the mud drum blowdown open, as this could drain the boiler and cause a unit trip due to low water level. Excessive blowdown also results in loss of energy.

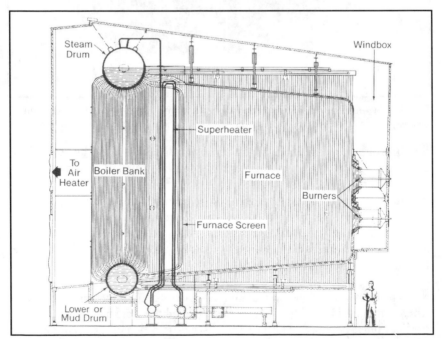

**Fig. 1-5:** Outline of a small industrial boiler (Reproduced with permission from *Steam*, 40th ed., published by Babcock & Wilcox, a McDermott Company)

# Steam generating circuits

Figure 1-6 outlines the component arrangement of a large utility boiler. Although the design is rather complex, it is being introduced now because it illustrates most of the important pieces of equipment in a steam generator. This section discusses water and steam circuit configurations, which will be helpful in the examination of common boiler designs later in the chapter.

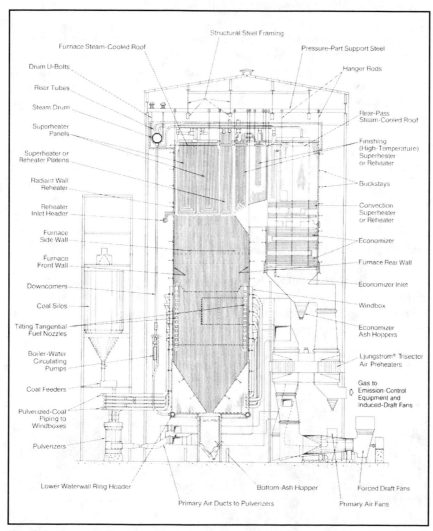

**Fig. 1-6:** Illustration of a large, subcritical boiler (Reproduced with permission from *Combustion: Fossil Power*, Alstom)

# Waterwall tubes

Most large drum boilers have a vertical waterwall tube arrangement, although a horizontal orientation is common in certain types of boilers or in specific areas of boilers. Floor tubes in cyclone boilers and arch tubes in boilers of many types are two examples of non-vertical tubing. Vertical design allows the tubes to be suspended from the boiler ceiling, which in turn eases stress on the tubes as they expand and contract at start-ups and shutdowns.

The most common tube design is straight-wall, although alternative designs have been developed to enhance turbulent flow and uniform steam generation within the tubes. One of the most popular alternative designs is the ribbed tube, which contains a spiraled pattern of raised ridges with geometry similar to the pattern in a rifle barrel. The grooves impart turbulent characteristics to the water that helps ensure uniform boiling.

The waterwall concept allows boilers to be designed with very little refractory material, as the water flowing through the tubes keeps them cool and prevents thermal failure. The mean tube temperature in a properly operated boiler is around 800°F (427°C), which is suitable for mild carbon steel, the preferred material for waterwall tubes. Factors that may cause temperature excursions in waterwall tubes include direct flame impingement on the tubes and, more commonly, waterside buildup of scale and iron oxide deposits inhibiting heat transfer.

The common structural configuration of waterwall tubes is illustrated in Figure 1-7. This is known as a membrane design. Construction of a large boiler usually involves fabrication of numerous waterwall membrane panels, which are

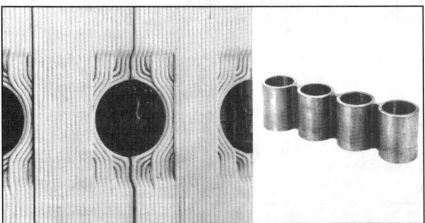

**Fig. 1-7:** Waterwall tubes showing membrane construction (Reproduced with permission from *Steam,* 40th ed., published by Babcock & Wilcox, a McDermott Company)

field erected. Feed to the waterwall circuits from the downcomers enters through headers at the base of each waterwall circuit, *i.e.,* front wall, sidewall, rear wall, etc. The headers are usually interconnected to help distribute flow evenly throughout the boiler.

The membrane design, with an exterior coating of insulation, confines heat to the boiler and provides for high heat transfer to the water/steam fluid in the tubes. In coal-fired boilers, the tubes accumulate slag (the molten residue of mineral matter, see chapter 3), which inhibits heat transfer. Waterwall tubes in these units are usually designed with studs extending outwards towards the furnace. The studs increase heat transfer, and are especially helpful in transferring heat when the tubes are coated with slag.

Waterwall tubes in natural circulation units are typically 2" to 4" in diameter (50.8 to 101.6 mm), while those in forced-circulation units may only be an inch (25.4 mm) or so in diameter. Diameters are larger in natural-circulation units to reduce frictional losses. The advantage of smaller tubes in forced-circulation units is wall strength. It is possible to construct thinner walls, which keep the tubes cooler. Forced circulation also improves the uniformity of flow through all of the circuits. An important design feature in forced-circulation units is that the tubes are orificed at the lower headers. This is done to ensure uniform pressure drop and flow through each tube so that some do not become starved of fluid and overheat. The effect is less pronounced in natural-circulation units, so orificed tubes are not generally needed. An interesting need for temporary orificing of natural-circulation boilers occurs during chelant chemical cleaning, where the solution must be heated to enhance flow through the tubes and to increase its reactive potential. The downcomers must be orificed to prevent short-circuiting of the chemical around the waterwall tubes.

## Superheaters and reheaters

Steam exits the drum in a saturated state. Saturated steam is not efficient for turbine operation, as the temperature drop through the turbine would cause the steam to condense and be of no value. A significant improvement to boiler efficiency came with the development of the superheater.

Superheaters are a series of tubes placed within the flue gas path of the boiler, whose purpose is to heat the boiler steam beyond saturated conditions. The two general categories of superheaters are radiant and convective, and both types are illustrated in Figure 1-6. Radiant superheaters are, as their name implies, exposed to radiant energy in the furnace, while convection superheaters sit further back in the gas passage and are shielded from radiant heat. The radiant superheater shown in Figure 1-6 hangs as a pendant section within the boiler. This configuration is also common for convective superheaters in the horizontal region of the flue gas duct. Superheater circuits are sometimes embedded within the upper waterwall tubes, and quite often superheater and reheater loops are located further along in the backpass of the boiler, just ahead of the economizer and air heater.

Superheaters are typically split into two sections—primary and finishing. The primary superheater is first in the network, and the finishing superheater completes the heat transfer process, bringing the steam to the temperature required by the turbine. The arrangement varies from boiler to boiler. Some boilers have only a small amount of superheat area exposed to radiant heat; in other cases, radiant superheaters are the finishing superheaters.

The increase in temperature to which steam is raised above the saturation point is known as the degree of superheat. Consider again a 2,400 psig (16.54 mPa) boiler. Steam tables show that the saturation temperature at this pressure is 663°F (350°C). If the steam is heated to 1,000°F (538°C) for use in a turbine, then it has 337°F (188°C) of superheat. Modern utility boilers typically have final steam temperatures of 1,000°F to 1,005°F (541°C), although some units have been designed with final steam temperatures of 1,050°F (566°C) and on a few occasions, 1,100°F (593°C). Higher steam temperatures are rare due to material performance issues.

Ash fouling of superheater tubes (chapter 3) is a major concern during the design and operation of a boiler. Fouling potential is greatest in the hottest portions of the convection pass, so wider spacing between superheater tubes is required in these areas to prevent bridging of ash deposits. Figure 1-8 illustrates the proportional spacing of superheater tubes as a function of flue gas temperature. Tighter tube bundles are possible further along in the flue gas path, where fouling potential is lower.

Reheat is a design modification to steam generating units that improves efficiency, and is standard with large boilers. The general steam flow path in a unit with a single-reheat loop is illustrated in Figure 1-9. The reheater increases temperature, not pressure, but the temperature gain still improves efficiency. Common designations for boilers list both the superheat and reheat temperatures. A boiler spoken of as "1,000°F/1,000°F" has identical superheat and reheat temperatures. Like superheaters, reheaters may be placed at various locations within the gas path.

Control of steam temperature in the superheater is important to maximize efficiency and prevent overheating of tubes. The common method of steam temperature control is attemperation from a spray of feedwater introduced directly into the steam. The most common feed point is between the primary and secondary

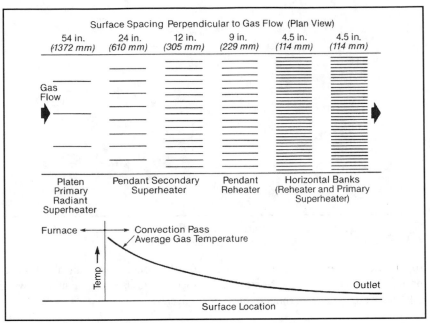

**Fig. 1-8:** Representative superheater/reheater spacing as a function of temperature (Reproduced with permission from *Steam*, 40th ed., published by Babcock & Wilcox, a McDermott Company)

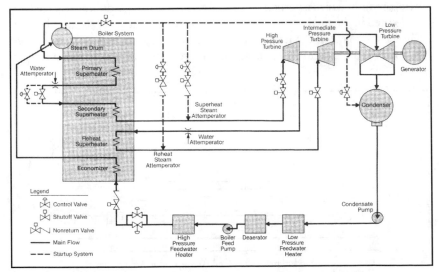

**Fig. 1-9:** General water and steam flow schematic of a drum boiler. Included is a start-up steam network (Reproduced with permission from *Steam*, 40th ed., published by Babcock & Wilcox, a McDermott Company)

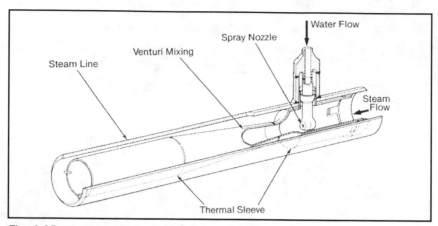

**Fig. 1-10:** Illustration of an attemperator spray nozzle (Reproduced with permission from *Steam*, 40th ed., published by Babcock & Wilcox, a McDermott Company)

superheaters. Feed after the secondary superheater could potentially allow water droplets to enter the turbine. Some high-temperature units are also equipped with reheater attemperators, although this is not universal.

Direct attemperation requires a specialty nozzle (Fig. 1-10). Nozzle design and materials minimize the effects of thermal stress when the relatively cool feed-water enters the hot steam line. Another type of attemperator found in older two-drum units is the cooling coil attemperator (Fig. 1-11). This is a non-contact attemperator in which a portion of the main steam is bypassed through tube bundles located in the mud drum and then reintroduced to the main steam. The cooler boiler water lowers the temperature of the bypass steam. Control valves adjust the flow rate of steam to the attemperator in accordance with main steam temperature.

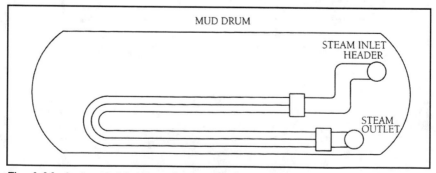

**Fig. 1-11:** Outline of a Mud Drum Cooling Coil Attemporator Arrangement

Figure 1-9 also illustrates a piping arrangement to assist with unit start-up. The key feature is the steam bypass directly to the condenser. As a unit starts, the steam has a tendency to be wet at first. This is not ideal for the turbine. An initial bypass of steam to the condenser allows the unit to develop some heat before the steam is introduced to the turbine. Many plants do not have this option, however, and must introduce wet steam to the turbine at start-up.

## Economizer

Toward the further reaches of the convection pass sits the economizer. As with reheaters, most modern, large steam generating boilers have an economizer. The economizer extracts additional heat from the flue gas, but in this case transfers the energy to the feedwater. Figure 1-12 outlines a typical economizer arrangement. Economizer tubes may be finned to provide additional heat transfer, although the fins increase the potential for fly ash accumulation.

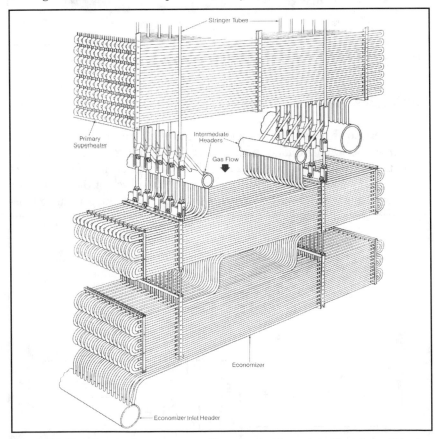

**Fig. 1-12:** Typical economizer arrangement (Reproduced with permission from *Steam*, 40th ed., published by Babcock & Wilcox, a McDermott Company)

**Fig. 1-13:** Cutaway view of a D-type boiler (Reproduced with permission from *Steam*, 40th ed., published by Babcock & Wilcox, a McDermott Company)

## Boiler designs

This section begins with a quick look at natural gas-fired package boilers and then progresses through utility boiler design from small to large. Figure 1-13 shows the very common D-type package boiler design. Firing is from the front of the boiler. The waterwall tubes surround the sides and back of the combustion chamber. This is a two-drum boiler with a separate set of steam generating tubes

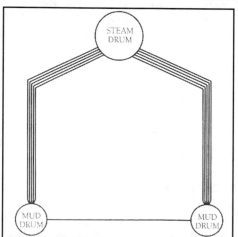
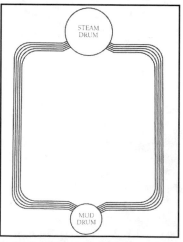

**Figs. 1-14 and 1-15:** General circuitry of an A–type boiler (left) and an O-type boiler (right)

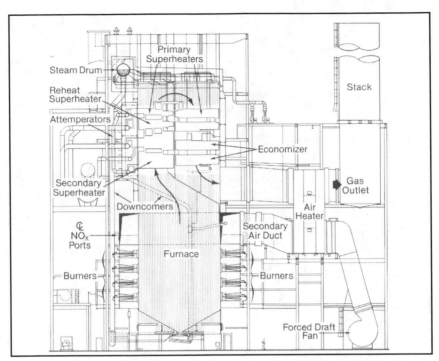

**Fig. 1-16:** The natural-gas fired El Paso-type boiler (Reproduced with permission from *Steam*, 40th ed., published by Babcock & Wilcox, a McDermott Company)

that directly connect the two drums. Known as the boiler bank, they are a common feature on low- and intermediate-pressure boilers to increase waterwall heat transfer. Although small in size, this boiler also has a superheater.

Two other designs are common for package boilers. These are the "A" and "O" types, whose basic outlines are shown in Figures 1-14 and 1-15. Operation is very similar to the "D" type.

The obvious advantage of package boilers is that they can be factory assembled and shipped to the job site in one piece. Common steaming rates in package boilers range from a few thousand pounds per hour to 100,000 pounds per hour. For larger applications, particularly at utilities, field-erected units are the norm. Before the proliferation of simple and combined-cycle combustion turbine units, a popular natural gas-fired boiler for utility applications was the El Paso type shown in Figure 1-16. This boiler is capable of producing almost 4 million pounds of steam per hour. An interesting feature of this unit is that the secondary superheater is first in the gas path, followed by the reheater, then the primary superheater. This is an opposed-fired boiler where the two sets of burners face each other from across the furnace.

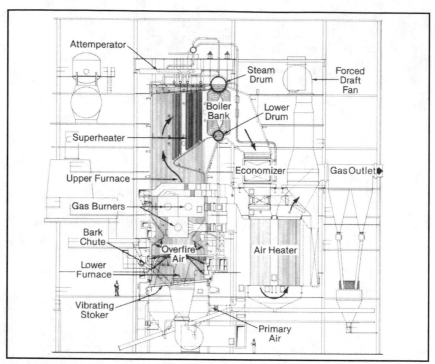

**Fig. 1-17:** Stirling™ power boiler (Reproduced with permission from *Steam*, 40th ed., published by Babcock & Wilcox, a McDermott Company)

A popular design for intermediate pressure applications is the Stirling™ power boiler (Fig. 1-17) named after the inventor, Allen Stirling. This is a two-drum boiler with a boiler bank to increase heat transfer. These types of boilers can be designed to fire a variety of fuels. The figure shows a unit with a vibrating stoker that can fire wood bark. (Stoker boilers can fire many fuels, as we shall examine in chapter 2.) This modern version of the Stirling boiler also has overfire air ports and supplemental natural gas burners for pollution control and flexibility of operation.

The Stirling™ boiler illustration clearly indicates furnace nose tubes. These are the tubes that angle out from the rear wall just below the superheater pendants. Nose tubes shield the superheater from radiant heat and help to prevent overheating of superheater tubes. However, the change from vertical alignment to a partial horizontal direction influences fluid conditions within the tubes. One potential effect is steam/water separation, with the steam flowing at the top of the tubes. This can cause overheating. Nose tubes are also notoriously susceptible to deposit buildups. Even when the rest of the boiler is clean, deposits in nose tubes may build up to levels that could cause corrosion or creep due to overheating. This is a prime area to collect tube samples for chemical cleaning deposit analyses.

Figure 1-18 shows a popular boiler from the 1960s and early 1970s—the cyclone. The main feature is the cyclone combustion chamber, of which most cyclone boilers have multiple units. The popularity of the boiler stemmed from good combustion efficiency. Pebble-sized coal is fed into the cyclones with air inlets at various locations, including along the length of the combustion barrel. The air imparts a swirling motion to the coal, hence the cyclone name. The barrels are water-cooled just as in a regular boiler, although the tubes in the barrel are typically around 1" in diameter. The tubes extend from an upper header at the top of the barrel and then connect into a lower header at the bottom of the barrel. The small diameter and curved configuration of the barrel tubes makes them somewhat

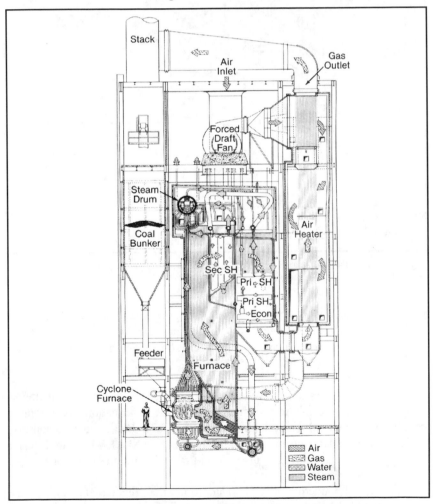

**Fig. 1-18:** Cyclone boiler (Reproduced with permission from *Steam*, 40th ed., published by Babcock & Wilcox, a McDermott Company)

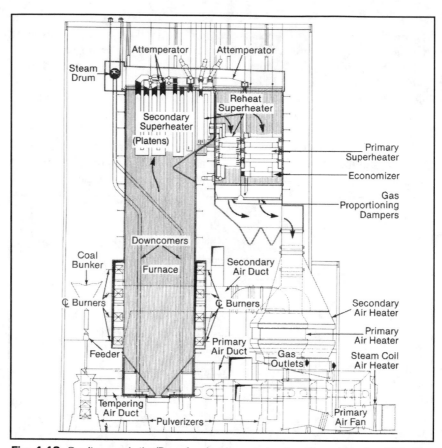

**Fig. 1-19:** Carolina-type boiler (Reproduced with permission from *Steam*, 40th ed., published by Babcock & Wilcox, a McDermott Company)

susceptible to deposit accumulation due to low flow. The remainder of the boiler has similar features to those already outlined, including vertical waterwall tubes and pendant superheaters and reheaters. Cyclone burners with a Stirling™ furnace are a common arrangement.

An important feature of cyclone boilers is that residual mineral matter is designed to exit the combustors as molten slag. (More information on this topic is covered in chapter 3.) These boilers fell out of favor because the combustion process generates large quantities of nitrogen oxides ($NO_x$), which are a major atmospheric pollutant and a precursor to acid rain and ground-level ozone.

Virtually all modern coal-fired boilers (other than fluidized-bed units) burn pulverized coal. A popular design is the tangentially fired unit, in which the water-wall configuration is box-shaped with burners located at several levels along the

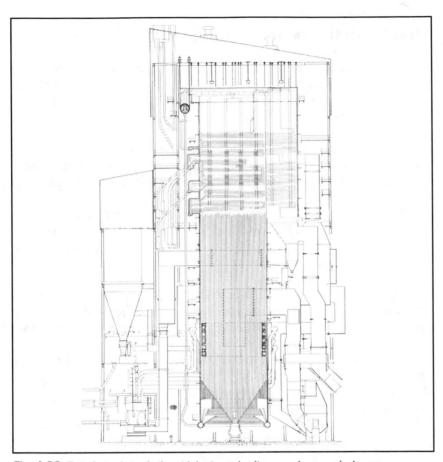

**Fig. 1-20:** Forced-circulation boiler with horizontal radiant superheater and reheater
(Reproduced with permission from *Combustion: Fossil Power*, Alstom)

corner of the furnace. This boiler was previously illustrated in Figure 1-6. The burner arrangement establishes a swirling fireball within the center of the furnace. This design is popular for large utility boilers. One important aspect of these units is that the burners may be tilted upwards or downward to alter the balance of heat transfer between the waterwall tubes and superheater/reheater. This is especially useful for handling load swings.

Another popular design is the Carolina-type boiler (Fig. 1-19). This boiler also fires pulverized coal. The unit illustrated in the figure is of the opposed wall firing type, in which burners are placed directly across from each other. The flames meet in the center of the furnace.

Yet another boiler design is outlined in Figure 1-20. This is a forced-circulation unit whose main features are horizontal superheaters and reheaters.

# Heat absorption patterns within boilers

Different boilers have different heat absorption patterns. This, of course, is dictated by boiler configuration, combustion patterns, heat exchanger locations, and other factors. Figure 1-21 illustrates just a few of these variations, but they give a good idea of typical patterns.

Boiler "a" is a simple industrial boiler designed to produce saturated steam at 200 psig (1.48 mPa). This is one of the low-heat boilers that has a boiler bank to improve steam capacity. The boiler bank takes nearly half of the heat input.

Now examine boiler "b." Similar to boiler "a," it is configured to produce superheated steam at 750°F (399°C) and 600 psig (4.24 mPa). Waterwall heat absorption does not change much, but more than 10% of the fuel energy is transferred in the superheater.

Comparing boilers "c" and "d" with "a" and "b"—and with each other—reveals several interesting details. First, an increase in superheater pressure and temperature often does not directly correlate with a percentage energy transfer decrease in the waterwall tubes. As is evident, the percentage of energy transfer in the waterwall tubes of "c" and "d" is greater than in the first two boilers. However, the size and heat absorption capacity of the superheater in boiler "c" is considerably greater than that in boiler "b." This stands to reason, as the superheater outlet temperature has been increased from 750°F to 1,005°F (541°C). Another interesting feature comes in comparing the boiler bank and economizer in "c" and "d," respectively. It

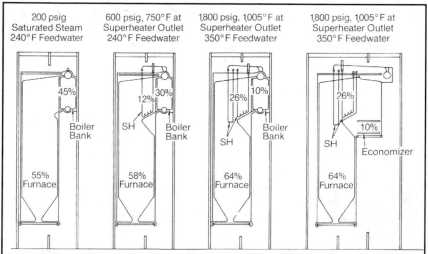

**Fig. 1-21:** Heat absorption patterns for four pulverized coal boilers (Reproduced with permission from *Combustion: Fossil Power*, Alstom)

is the author's experience that boilers at this pressure (and higher) typically have an economizer and not a boiler bank. Regardless, each accounts for about 10% of heat transfer in the steam generator.

## Once-through and supercritical boilers

An important class of boilers is the once-through type (Fig. 1-22), most of which are used in supercritical applications. These boilers are also known as universal pressure steam generators. The chief feature of once-through boilers is that

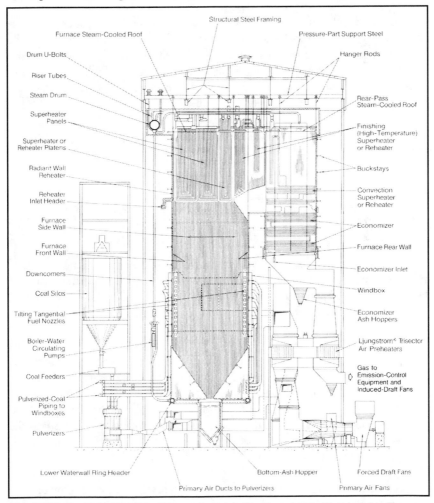

**Fig. 1-22:** Combined Circulation™ once-through boiler (Reproduced with permission from *Combustion: Fossil Power*, Alstom)

all of the incoming water to the boiler is converted to steam in the waterwall tubes. The steam is collected in headers for transport to the superheater. The advantage of supercritical operation is better efficiency. Some units operate at up to 4,500 psig (31.1 mPa) pressure.

An important factor with regard to once-through operation is water purity. The conversion of all the incoming feedwater to steam allows direct carryover of contaminants to the superheater and turbine. It is absolutely imperative that only the highest quality feedwater enter the boiler. This mandates that these units be equipped with condensate polishers.

A question that naturally comes to mind is how does one start up a once-through unit? At start-up, the boiler is not nearly hot enough to convert all of the water to steam. Obviously, water cannot be allowed to enter the superheater and turbine. In the combined-circulation unit, water exiting the boiler tubes at unit start-up flows to a receiving vessel from which it is routed through a downcomer to lower waterwall headers. The boiler somewhat resembles a drum unit at start-up, with the notable exception that the water-receiving vessel does not have a steam outlet. Once the boiler has reached minimum operating temperature, the recirculation loop is isolated.

A common tube design for modern supercritical units is the spiral-wound type, in which the waterwall tubes angle along the furnace walls, and thus have both a vertical and horizontal component. This design has been shown to improve heat transfer efficiency. The Foster Wheeler Corporation has been heavily involved in design and development of these units. One drawback of the spiral-wound design is that the tubes cannot be suspended from the ceiling as vertical tubes can, requiring special tube supports along the furnace walls. Extra support brackets are always potential sites for fatigue-related tube failures. Frictional losses are also higher in spiral tubes.

Another type of popular supercritical unit incorporates a combined spiral-wound vertical waterwall tube design. This is the Benson-type popularized by Babcock Borsig Power. The lower portion of the furnace has spiral-wound tubes that tie into vertical tubes through headers higher up in the boiler. An interesting concept is being promoted with installation of these units—where the plant also has a hyperbolic, mechanical-draft cooling tower—to route the flue gas into the cooling tower above the film-fill packing. This increases buoyancy in the tower, and combines two plumes into one.

Supercritical steam generation offers potential for the next generation of coal-fired power plants. A basic law of thermodynamics states that the maximum efficiency of a heat engine is described by the equation:

$$\eta = 1 - T_L/T_H$$

*where:*

$\eta$ is the efficiency

$T_L$ is the low temperature heat sink (steam condenser)

$T_H$ is the high temperature source (boiler)

Advances in supercritical design and materials technology will lead to increases in $T_H$ and improved efficiency. Even now, supercritical designs exist that favorably compare with subcritical drum units in terms of capital and cost. Fuel savings may quickly offset the slight additional capital cost of the supercritical unit. Units with thermal efficiencies at or near 45% are becoming common. This is with superheat and reheat steam temperatures in the 1,150°F to 1,180°F (621°C to 638°C) range.

## Conclusion

This chapter outlined fundamental water/steam details of many conventional fossil-fired boilers. A wide variety exists. Chapter 2 examines some of the more modern steam generating systems including fluidized bed combustors, heat recovery steam generators, and a venerable combustion technique—stoker firing—that is being used to burn alternative fuels such as trash, wood, chipped tires, and other non-traditional fuels.

# Appendix 1-1

Figure A1-1 shows a common condensate/feedwater scheme for a utility boiler. The condensate flows through a series of tube-in-shell, low-pressure feedwater heaters to the deaerator, which is an open feedwater heater designed to remove dissolved oxygen and other gases from the condensate. The feedwater pump pressurizes the condensate for injection into the boiler. Before entering the boiler, the feedwater passes through a series of high-pressure heaters and the economizer. The heating source for the feedwater heaters is extraction steam taken from various points in the turbine, while the economizer sits in the flue gas path.

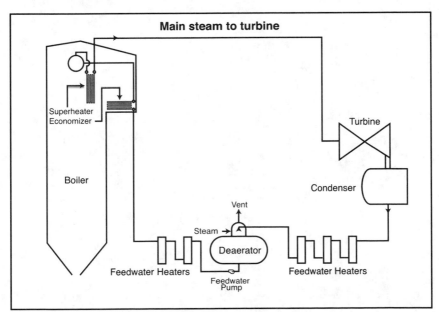

**Fig. A1-1:** Simplified utility water/steam network showing feedwater heaters.

Condensate/feedwater heating improves boiler efficiency by utilizing some of the heat that would otherwise be lost in the condenser. At some point, a balance is reached between efficiency gain and capital cost for the heaters. Six heaters are very common for sub-critical utility boilers and the author has heard of as many as eight, including the deaerator.

In the deaerator, the incoming condensate is mixed with live steam in a compartment that may include trays to enhance water/steam contact. The mixing process and increased temperature liberate dissolved gases that might otherwise cause corrosion in the boiler. The heated condensate flows downward to the deaerator storage tank, while the liberated gases and a small amount of steam are vented from the top of the unit.

# Appendix 1-2

Waterwall tubes depend upon the flow of water along the inside tube surface to maintain steady temperatures. Obviously, as water rises from the bottom to the top of the tubes, the water temperature increases and more steam is created. In a properly designed steam generator, however, the boiling process is uniform and is known as nucleate boiling.

Upsets in burner operation, rapid load increases, poor design, and other factors will affect the boiling process and may induce a "departure from nucleate boiling" (DNB) that alters the cooling effects of the boiler water. As Figure A1-2 illustrates, DNB reduces heat transfer to the boiler water. This in turn places the tubes in the affected areas under higher temperatures than originally designed. "Creep" may then set in with eventual tube failure as the result. Internal tube deposits also restrict heat transfer and may cause similar tube failures.

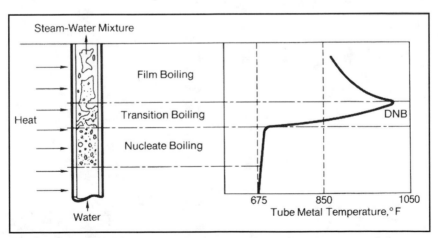

**Fig. A1-2:** Effect of DNB on tube metal temperature (Reproduced with permission from *Combustion: Fossil Power*, Alstom)

# Appendix 1-3

Sodium phosphate compounds [principally tri-sodium phosphate ($Na_3PO_4$), or TSP, as it is commonly known] have been popular for boiler water treatment for years. The chemical serves two primary purposes. First, it reacts with water to produce an alkaline solution:

$$Na_3PO_4 + H_2O \rightarrow NaOH + Na_2HPO_4$$

Corrosion of mild steel is minimized within a pH range of 9 to 11, which is where phosphate programs can be adjusted.

Second, phosphate reacts with contaminants (in particular, the scale-formers calcium, magnesium, and silica) that enter the boiler via condenser tube leaks or other sources. Phosphate reacts directly with calcium to produce calcium hydroxyapatite:

$$10Ca^{+2} + 6PO_4^{-3} + 2OH^- \rightarrow 3Ca_3(PO_4)_2 \cdot Ca(OH)_2 \downarrow$$

Magnesium and silica react with the alkalinity produced by phosphate to form the non-adherent sludge, serpentine:

$$3Mg^{+2} + 2SiO_3^- + 2OH^- + H_2O \rightarrow 3MgO \cdot 2SiO_2 \cdot 2H_2O \downarrow$$

These "soft materials" are much easier to remove than the hard scale or corrosive compounds that would otherwise form. The mud drum serves as a collection point for these soft sludges. Periodic manual blowdown is necessary to prevent these materials from building up to unwanted proportions.

# Chapter 2

## The "Newer" Technologies– Fluidized–Bed Combustion, Combined–Cycle Power Generation, Alternative Fuel Power Production, and Coal Gasification

Although the technologies listed in this chapter might sound new, most have been around for a long time. What's driving them are environmental and efficiency issues. As they have become increasingly critical factors, these technologies have taken on much more importance. The natural gas-fired combined-cycle process is both efficient and environmentally friendly, while fluidized-bed combustion provides pollution control directly in the boiler and offers another option for continued use of coal, as does integrated coal gasification. Alternative fuel (refuse, biomass) boilers are often stoker fired, which has kept this venerable technology alive.

## FLUIDIZED BED COMBUSTION

As chapter 1 indicated, many of the most modern coal-fired units have been based around pulverized coal technologies. However, environmental regulations regarding nitrogen oxide ($NO_x$) and sulfur dioxide ($SO_2$) emissions require pollution control techniques. They can include burner modifications, overfire air, selective-catalytic-reduction (SCR), or selective non-catalytic reduction (SNCR) for $NO_x$ control, and fuel switching, wet scrubbing, or dry scrubbing for $SO_2$ removal. These pollution control methods add a great deal of complexity and cost to a steam-generating unit.

Fluidized-bed combustion incorporates $NO_x$ and $SO_2$ control directly into the furnace design and operation. The term "fluidized bed" refers to the process in which an upward flow of air keeps fuel and ash particles in a suspended state within the boiler. The airflow required to initially transform a fixed bed of material into a fluidized bed is known as the minimum fluidization velocity. This velocity is dependent upon a number of factors, including fuel type, fuel size, and air temperature. Some fluidized-bed boilers are designed to operate with airflows only marginally above the minimum. These are known as bubbling-fluidized-beds (BFB). The bed remains in a relatively compact layer within the furnace and air bubbles up through it. The other type is the circulating-fluidized-bed (CFB), in which higher air velocities circulate in the bed throughout the furnace. Figure 2-1 illustrates general details for both types of fluidized-bed boilers.

This discussion focuses on the circulating fluidized-bed and specifically on the atmospheric circulating fluidized-bed process (henceforth referred to as ACFB), as it is a popular technology for utility applications.

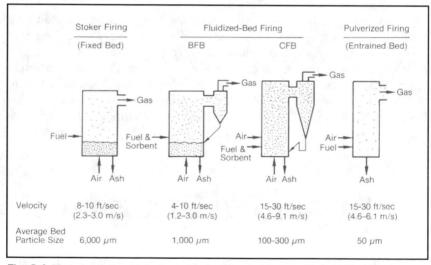

**Fig. 2-1:** Typical airflow, particle size, and bed volume data for several standard boilers (Reproduced with permission from *Combustion: Fossil Power*, published by Alstom)

# Design overview

ACFB offers several advantages that make the technology attractive. Of primary importance is the combination of relatively low combustion temperatures (1,500°F to 1,700°F, 816°C to 927°C) and surprisingly high heat transfer. In a conventional boiler (such as a pulverized-coal unit or a cyclone), flame temperatures may reach 3,000°F (1,649°C). Radiant heating is a major energy transfer mechanism, but these same high temperatures also generate significant quantities of nitrogen oxides (see chapter 5 for a more detailed discussion of this process). High heat transfer occurs in a CFB, but in this case, it is due to a bed of burning coal particles and hot ash that transfer heat in large part by convection to the boiler tubes as the bed moves through the furnace. The lower temperatures limit $NO_x$ production, while the bed dynamics are utilized to introduce chemical reagents that react with and remove $SO_2$.

Figure 2-2 shows a general flow schematic for a common ACFB design. The basic process and important features are as follows:

- Coal enters the boiler above an air distributor plate located at the bottom of the combustor

- The air fluidizes the coal to form an agitated bed of burning particles within the combustor. The combustor in and near the fuel zone is lined with refractory, as very abrasive conditions are present in this area

- Limestone (sorbent) injected to the combustor or added to the fuel reacts with $SO_2$ and oxygen to produce calcium sulfate ($CaSO_4$), a solid and benign waste product. This removes $SO_2$ directly in the furnace, and eliminates or minimizes the need for backend $SO_2$ control

- The combustion temperature is usually maintained at or close to 1,550°F (843°C). The low temperature minimizes $NO_x$ formation. Staged air feed is another process that reduces $NO_x$ production

- The relatively low combustion temperatures are below ash fusion temperatures for virtually all coals. This eliminates boiler slagging and in turn provides flexibility in the types of fuel that may be burned

- The bed flows upward through the furnace and enters a mechanical cyclone that returns heavier unburned coal particles and most ash to the combustor. Flue gas and light ash particles flow from the top of the cyclone to the flyash handling system

- A bottom ash removal system helps to maintain bed inventory and keep the bed from overloading with material

- As with conventional boilers, steam is generated within waterwall tubes located in the furnace. Although refractory may cover the tube surfaces in the combustor, further up in the furnace the tubes are directly exposed to the environment, as is typical for a coal-fired boiler. Superheater pendants hang in the convective path similarly to those in other boilers. Some CFB designs include heat exchangers, shown in Figure 2-2 in the fluidized-bed heat exchanger (FBHE), the cyclone return to the combustor and in the bed ash removal system. These are water or steam-cooled, and are incorporated into one or more of the boiler, superheater, or reheater networks.

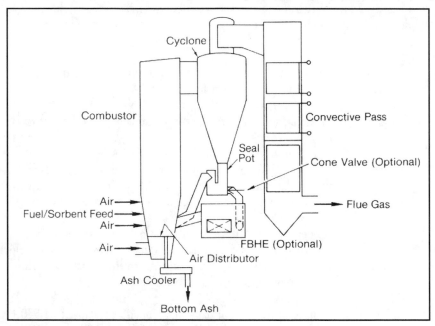

**Fig. 2-2:** Schematic of a common CFB boiler (Reproduced with permission from *Combustion: Fossil Power*, published by Alstom)

Coal may be introduced to the CFB by one of several methods. Pneumatic feeding at several ports in the combustor is one possibility. Another technique is feed from a chain conveyor that traverses the furnace. Air agitation from below mixes the fuel.

Fuel type influences coal preparation requirements. Coals that ignite easily may only need crushing similar to that for a cyclone boiler. Recommended top sizes for lignite and bituminous coal are 3/8" and 1/4", respectively. A low-volatile material such as anthracite culm has to be smaller in size. High-moisture coals like lignite may require drying to prevent clogging of coal feed lines. Air-swept crushers are one possibility for drying the fuel; another is to feed the fuel into the cyclone return line to the combustor. The hot gas stream dries the coal during its passage to the combustor.

Unit start-up requires gas or fuel oil for preheating. The burners reside either within the lower combustor or the cyclone return line to the boiler. These bring the boiler temperature up to around 1,000°F (538°C), at which point the combustor has become hot enough for the primary fuel to be introduced. Sand or alumina serves as a starter bed material until enough ash has circulated through the system to establish a natural bed.

Limestone for $SO_2$ removal may be added with the coal or blown into the unit at a short distance above the coal feed level. The vigorous agitation that occurs within a fluidized bed enhances the reaction between the limestone reagent and $SO_2$, and a well-operated CFB will remove 90% to 95% of the $SO_2$ produced during combustion. The common parameter for measuring the limestone reaction efficiency is the calcium/sulfur ratio. The basic reactions that occur between limestone and $SO_2$ are as follows:

$$CaCO_3 + heat \rightarrow CaO + CO_2 \uparrow$$

$$CaO + SO_2 + \tfrac{1}{2}O_2 \rightarrow CaSO_4 + heat$$

The first reaction absorbs heat (around 766 Btu/lb, 1,782 KJ/kg of $CaCO_3$ injected) and the second gives off heat (around 6,733 Btu/lb, 15,660 KJ/kg of sulfur removed). Theoretically, one equivalent amount of calcium reacts with one equivalent of $SO_2$ but as with all chemical processes, 100% efficiency is virtually impossible to achieve. Factors that influence the limestone-$SO_2$ reaction include the kinetics of sorbent particle-$SO_2$ interaction and the physical properties of the particles themselves. Sulfur dioxide tends to react at the surface of the sorbent particles, which can blind the interior of the particle from reaction. The type of limestone influences how well it calcines (converts to CaO). Particles that develop large pores during calcining tend to react more efficiently with $SO_2$. It is possible to measure the efficiency of the sulfur capture process by determining the ratio of calcium to sulfur in the byproduct. This value is known as the calcium/sulfur or Ca/S ratio. A common objective is to minimize the calcium to sulfur ratio, as the following discussion helps illustrate.

Note: Some in industry suggest that proper Ca/S ratios should average 1.1-1.2 to 1, but this efficiency seems almost too good to be true, at least in my experience. I have seen 1.5/1 listed in the literature as about the best to be expected, and this is the value outlined in the example illustrated below.

Reagent preparation is important. The maximum top size of limestone is reported to be 1 millimeter (1,000 microns). Grind sizes as low as 70% passing through a 200-mesh screen (127 microns) may be necessary. A fine grind size provides a large surface area for reaction, but if the limestone grind is too fine, unreacted particles may exit through the cyclone overflow. With the proper limestone quality and grind size, good limestone efficiency is possible. This was clearly demonstrated in a series of tests performed on the twin ACFBs (Units 6 and 7) at the University of North Carolina-Chapel Hill (UNC-CH) co-generation facility in the 1990s. Although the original limestone reagent had a $CaCO_3$ content ranging from 85% to 90%, reactivity was low, and at times the ash removed from the system contained up to 50% calcium oxide. The utility tested and subsequently switched to a limestone containing 95% $CaCO_3$. They also improved operation of the ball mills to optimize limestone particle size. The combination of these two factors greatly improved reaction efficiency to the point where unused reagent in the ash dropped to as low as 3%, and Ca/S ratios dropped from an average of almost 4.5/1 to less than 1.5/1.

The ideal temperature for $SO_2$ removal is 1,550°F, and reactivity can drop off significantly beyond this point. This is one of the major factors that require CFB temperatures to be held below 1,700°F.

Primary air to the combustor enters through a distributor plate located at the bottom of the furnace. This ensures an even distribution of flow to the bed. Airflow rates are critical and must be designed to fluidize the bed, but not push excessive material out of the boiler, as heat transfer in the furnace is greatly dependent upon hot particles contacting the tube walls. When coal particles burn, they become lighter in weight and rise. Counterbalancing this effect are complex interactions of particles, which tend to increase particle weight. All factors considered, airflow rates are generally higher than those that would be mathematically predicted for individual particles. As Figure 2-1 illustrates, the linear airflow rate for an ACFB boiler ranges from 15 to 30 feet per second (4.6 to 9.1 meters per second).

Direct flames are not as evident in a CFB as they are in a pulverized coal or cyclone unit. Rather, the process resembles a swirling sandstorm with an orange glow. Temperature control is very critical with regard to NOx production. During the limestone tests at UNC-CH, the bed temperatures increased by 60°F to 70°F (33°C to 39°C) over the normal averages of 1,640°F (893°C) and 1,650°F (899°C) in Units 6 and 7, respectively. Corresponding $NO_x$ emissions on Unit 6 increased by 87% and on Unit 7 increased by 200%. This illustrates the criticality of maintaining combustion temperatures at design levels.

Note: The bed temperature increases were theorized to be due to the increased reactivity of the new limestone. This effect is explained in Appendix 2-1.

Air staging is another another $NO_x$ control technique. Primary air feed is held below stoichiometric levels, and the remaining air is added above the primary combustion zone. Chapter 5 discusses in detail how this lowers $NO_x$. A properly designed and operated CFB can reduce $NO_x$ emissions to 0.2 pounds per $10^6$ Btu (0.085 kg/$10^6$ kJ), and emission rates as low as 0.1 lb/$10^6$ Btu (0.043 kg/$10^6$ kJ).

The low-combustion temperatures in a CFB also prevent slagging. Ash fusion temperatures of most coals are typically above 2,000°F (1,093°C) even in a reducing atmosphere. This is at least 300°F (166°C) above the top temperature in the fluidized bed. Fouling of superheaters and convective pass structures can still be a problem, as furnace temperatures are still high enough for sodium to form the usual vaporous oxide (see chapter 3).

A principal advantage of CFB units centers around fuel residence time. In a pulvarized coal unit, the residence time may only be two or three seconds; residence time in a CFB theoretically could be infinite, as the design forces heavier, unburned particles to return to the furnace. A CFB has high-combustion efficiencies due to the circulation process and the fact that unburned carbon is returned to the combustor. The cyclone separator operates similarly to its counterparts in flue gas desulfurization systems and other industrial processes. By imparting circular motion to the gas, the cyclone causes a separation of particle size by weight. Light particles (typically of 100 microns in size or smaller) flow out of the top discharge port of the cyclone, while unreacted coal and lime, along with heavier ash particles, discharge through the cyclone underflow and return to the boiler.

The gas flow rate entering the cyclone usually must be above 100 feet per second to effect good particle separation. This causes severe wear in the cyclone, which is lined with refractory material. Even with these precautions, cyclones are one of the most maintenance-intensive pieces of equipment in the boiler. An alternative to the cyclone collector is shown in Figure 2-3, in which a series of U-beams force the gas through a torturous path, causing the bed particles to fall into the circulation system.

A common auxiliary system on larger units is a heat exchanger placed in the gas stream from the cyclone exhaust to the boiler. Depending upon unit design, the heat exchanger may be part of the boiler, superheater, or reheater network. A similar heat exchanger is sometimes used in the bed ash removal system, especially in larger units. Bed ash must be withdrawn to maintain bed density at a proper concentration. The bed ash system takes material from lower levels of the unit. The ash temperature is around 1,500°F (816°C), and must be cooled to a range of 200°F to 400°F (93°C to 204°C) before disposal. The heat exchanger is one method for cooling the ash. Another common method is to remove the ash through water-cooled screw conveyors and then send it to the ash removal system.

A prominent advantage of CFB combustion is fuel flexibility. Fuels may range from natural gas to low-rank coals to biomass. A generator can switch between

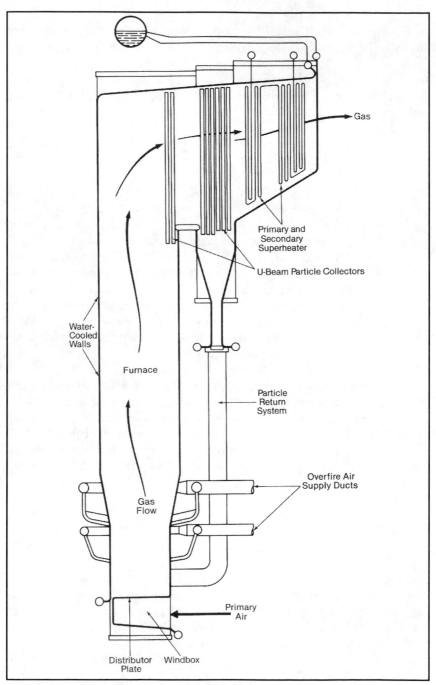

**Fig. 2-3:** CFB boiler with U-beam particle collectors (Reproduced with permission from *Steam*, 40th ed., published by Babcock & Wilcox, a McDermott Company)

coals without many of the major modifications that are necessary with other types of combustion. This allows plant managers to shop around for the most economical coal and change suppliers if necessary. At least one municipal electric utility seriously considered fluidized-bed combustion for a base-load generating unit because the CFB would allow the city to blend wood chips and leaves with the fuel. City ordinances prevent the burning of leaves and tree limbs within city limits, but a CFB would allow the city to derive an energy benefit from this free (transportation and collection labor excluded) biomass.

## Disadvantages of fluidized bed combustion

Like all technologies, fluidized bed combustion is not foolproof. Of prime concern is the corrosive character of the ash, especially when it is accelerated to high speed in the cyclone. Also, part of the furnace is lined with refractory—a material susceptible to thermal stress—that may spall off if the unit is frequently cycled. Thus, CFBs operate best in base-mode fashion. Fuel switching, while easier to do than in other types of boilers, may be complicated if the moisture content changes drastically.

Emissions regulations could present challenges for CFBs as well as other coal-fired boilers. A CFB does not remove mercury, for example, and so this pollutant has to be controlled by backend methods. Control of particulate matter less than 2.5 microns in diameter (PM2.5) is another regulation-in-progress and could complicate CFB design and operation. Emissions limits of 0.10 lb/$10^6$ Btu (0.045/$10^6$ kJ) for $NO_x$ and $SO_2$ may become reality for future units, and this could tax even the best CFBs. However, in-furnace removal of most of the $NO_x$ and $SO_2$ would greatly reduce load on any backend pollution control equipment, thus reducing capital and operating costs. CFBs already compare favorably with pulverized-coal fired units in terms of capital cost.

CFBs' versatility is illustrated in the Foster Wheeler advanced circulating fluidized bed plant, in which a pressurized circulating fluidized bed not only generates steam for power production but also generates a syngas that can be fed to a combustion turbine or co-generation process. Also under consideration are once-through supercritical CFBs. Like their sub-critical counterparts, they could offer a capital cost advantage over pulverized coal units that must be equipped with large, backend pollution control systems.

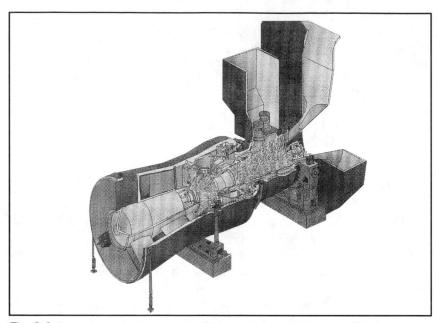

**Fig. 2-4:** Heavy-duty industrial gas turbine (Keihofer, Bachmann, Nielson, and Warner, *Combined-Cycle Gas & Steam Turbine Power Plants*, 2nd ed., PennWell Publishing)

## Combined-cycle power generation and heat recovery steam generators

A significant amount of electricity is now generated using simple-cycle and combined-cycle natural gas-fired (occasionally oil-fired) power plants. The basic outline of a simple-cycle gas turbine is shown in Figure 2-4. This process is technically referred to as the Brayton cycle. Compressed air produced in the front stage of the unit is fed to the combustor, where the burning fuel produces expanding gases that drive a vaned turbine, which, like a conventional steam turbine, spins a magnet inside a coiled generator to produce power. The process is not unlike that of a jet engine except that the turbine produces "shaft work" rather than thrust. A primary advantage of simple-cycle combustion turbines is that these units can be started quickly and are ideal for peak load situations. The disadvantage is that exhaust temperatures may be 800°F (426°C) or higher, which means that a great deal of energy leaves the system without being productively extracted. Also, the compressor consumes a great deal of energy. The maximum efficiency for a simple-cycle gas-fired generating plant is around 35%.

A very considerable efficiency improvement is possible by using the exhaust gas from the combustion turbine to heat a steam generator. When the steam is used for industrial purposes, the process is known as co-generation; when the steam drives a turbine for electricity production, the process is known as com-

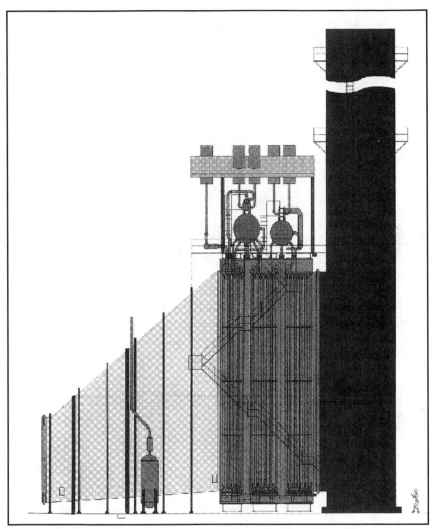

**Fig. 2-5:** Outline of a vertically-tubed, natural-circulation HRSG (Keihofer, Bachmann, Nielson, and Warner, *Combined-Cycle Gas & Steam Turbine Power Plants*, 2nd ed., PennWell Publishing.)

bined-cycle power generation. Efficiencies of combined-cycle units may approach 60%.

The key component in a combined-cycle plant is the heat recovery steam generator (HRSG). The primary modes of heating in HRSGs are convection and conduction, so the physical layout of waterwall, superheater, and reheater tubes is different than in conventional boilers. Figure 2-5 shows the general outline of the most common HRSG design, where the boiler tubes are vertically oriented and the gas turbine exhaust passes through in a horizontal direction.

Some HRSGs are single-pressure units, but much more common are multiple-pressure systems, as they offer improved efficiency. Figure 2-6 illustrates a triple-pressure HRSG—several aspects of this unit stand out. First is the configuration of the circuits within the HRSG. Those in the hottest zone are the high-pressure (HP) reheater, superheater, and evaporator (boiler). These are followed by the intermediate-pressure (IP) and low-pressure (LP) circuits. Each evaporator is set up in a natural circulation, drum-type arrangement, and the HP and LP circuits generate steam for power production. The LP circuit contains a combined drum/deaerator, which serves as the feedwater source for the IP and HP circuits. The deaerator is incorporated directly into the LP circuit to remove dissolved gases.

Combined-cycle power plants may be equipped with auxiliary, natural gas-fired duct burners to increase the heat flow to the HRSG, but this has not been universally adopted. In general—and especially in HRSGs without supplemental burners—heat fluxes are lower than in a conventional boiler, so the design often incorporates finned tubes to assist with heat transfer. The drawback to finned tubing is that the tubes are difficult to remove for deposit sampling or other maintenance activities.

Two important heat transfer criteria stand out regarding HRSGs—approach temperature and pinch-point temperature. The HRSG shown in Figure 2-6 has two economizers to heat feedwater. The approach temperature is the difference in

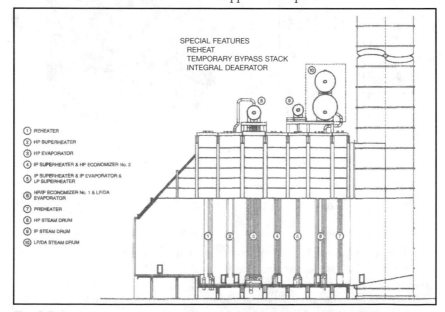

**Fig. 2-6:** Outline of a three-pressure HRSG (Illustration provided by Nooter Eriksen.)

temperature between the economizer outlet and the boiler circuit that the economizer serves. When factors such as economizer size, efficiency, and equipment
costs are taken into account, a properly designed economizer will bring the feedwater temperature to within 9°F to 22°F (5 to 12°C) of the boiler water saturation
temperature. Tighter approach temperatures would make the size and cost of the
economizer prohibitively large. In addition, narrow approach temperatures can
cause steaming in the economizer. Steaming may cause water hammer and fluctu-
ations in drum level.

The pinch-point temperature is the difference between evaporator steam outlet temperature and the exhaust gas temperature at that physical location in the
HRSG. A common range for pinch-point temperatures is 14 to 27°F (8 to 15°C).
Figure 2-7 illustrates by diagram the approach and pinch-point temperature rela-
tion for a single-pressure HRSG. Lower pinch-point temperatures require larger

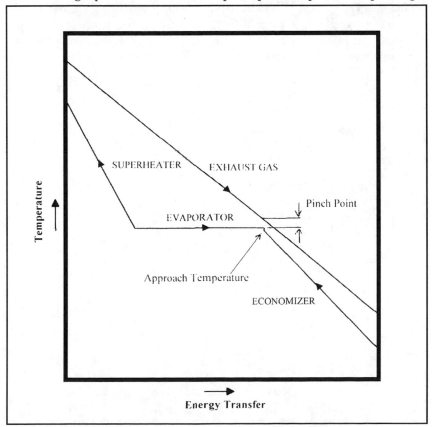

**Fig. 2-7**: Energy/temperature diagram for a single-pressure HRSG (Keihofer, Bachmann,
Nielson, and Warner, *Combined-Cycle Gas & Steam Turbine Power Plants,* 2nd ed., PennWell
Publishing)

heating surfaces, so the economic balance between efficiency and equipment cost dictates the optimum pinch-point temperature.

Many factors can influence the efficiency of a combined-cycle plant, and a discussion of most of these factors is beyond the scope of this book. (An excellent source for further reading is *Modern Power Systems*, by Pai and Engström. See reference section at end of book.) One concept that stands out, however, is that in a combined-cycle plant, the efficiencies of two systems (the gas turbine and the HRSG) influence the overall efficiency of the unit. In some cases, an efficiency decline in one positively affects the other! For example, a loss of efficiency in the gas turbine would increase the exhaust temperature, causing a rise in steam production in the HRSG. In general, efficiency losses in the gas turbine negatively affect overall performance, and it is best to design and operate the system with the gas turbine at maximum efficiency. Figure 2-8 shows the output and efficiency of a combined-cycle unit as HRSG back-pressure increases. Clearly illustrated is the decline in gas turbine output and efficiency. Increased back-pressure improves heat transfer in the HRSG, but the loss of efficiency in the gas turbine more than off-

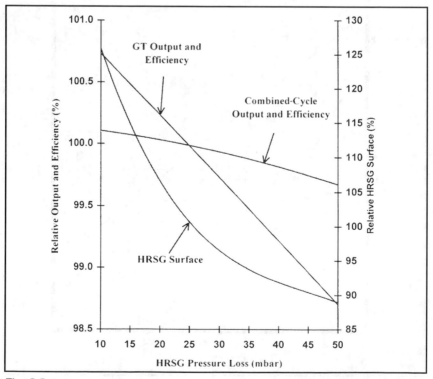

**Fig. 2-8:** Influence of HRSG back-pressure on combined-cycle output and efficiency, gas turbine output and efficiency, and HRSG surface (Keihofer, Bachmann, Nielson, and Warner, *Combined-Cycle Gas & Steam Turbine Power Plants*, 2nd ed., PennWell Publishing)

sets this gain, and the overall effect is to reduce the combined-cycle output and efficiency.

Not all HRSGs are of the vertical-tube drum design. Some units, particularly once-through HRSGs, may have a horizontal tube design (Fig. 2-9). These are not as common as the vertical-tube type.

One issue of primary concern in HRSGs is flow-assisted-corrosion. This corrosion mechanism, influenced by water chemistry, flow patterns, and temperature,

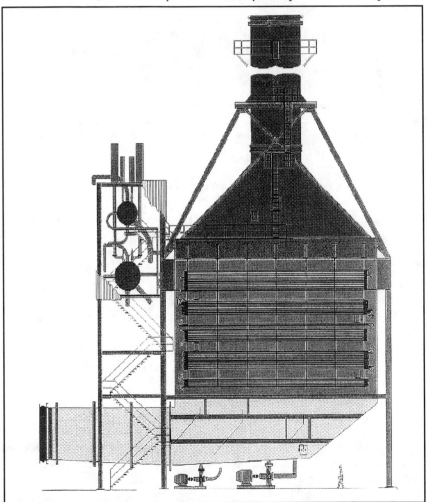

**Fig. 2-9:** Outline of a horizontally-tubed, forced circulation HRSG (Keihofer, Bachmann, Nielson, and Warner, *Combined-Cycle Gas & Steam Turbine Power Plants*, 2nd ed., PennWell Publishing)

has become quite a serious problem in many units. Appendix 2-2 provides additional details regarding this phenomenon.

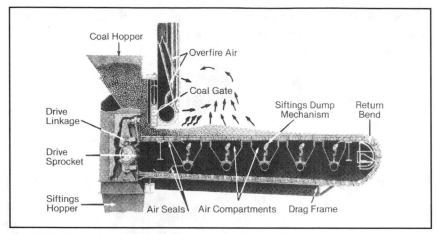

**Fig. 2-10:** Illustration of a chain grate stoker (Reproduced with permission from *Steam*, 40th ed., published by Babcock & Wilcox, a McDermott Company)

# Alternate fuel boilers and stoker-fired units

Stoker-fired boilers could have logically been introduced in chapter 1, as stoker firing has been around for a long time. However, stokers are not common for modern large generator applications, so the discussion has been delayed until now. Stokers are still popular for alternate fuel firing including biomass and refuse-derived fuel.

Stokers resemble other steam-generating units, with the notable exception that the fuel combusts on or just above a mechanical grate. Primary air feed for a stoker comes from below the grate, and fuel may be fed below the grate (underfeed) or above the grate (overfeed). Two of the most popular stoker types are the overfeed traveling grate stoker and the overfeed spreader stoker (depicted in Figs. 2-10 and 2-11, respectively). A discussion of how stokers operate with coal offers a good introduction into the topic of alternate fuel firing.

Consider first Figure 2-10, the outline of a traveling grate stoker. Coal is fed at the leading end of the stoker grate. Primary air feed comes from below the grate with some overfire air above the grate for $NO_x$ control. The underfire air prevents the grate from overheating. As the grate traverses the furnace, the coal combusts, being reduced to ash by the time it reaches the opposite side of the furnace. A variable underfire airflow pattern is common to provide more air to the coal at the entry point and less as the coal loses combustible material.

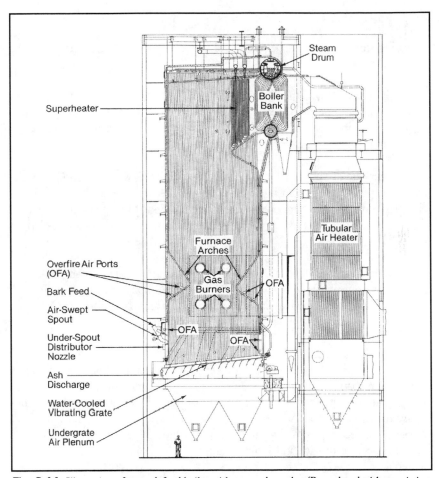

**Fig. 2-11:** Illustration of a wood-fired boiler with a spreader stoker (Reproduced with permission from *Steam*, 40th ed., published by Babcock & Wilcox, a McDermott Company)

Figure 2-11 shows a spreader stoker unit with a vibrating grate. In these boilers, mechanical spreaders distribute the fuel uniformly over the stoker grate to provide a more even combustion pattern. The vibration causes the fuel to flow to one end of the grate. With a properly designed furnace, the fuel is converted almost entirely to ash by the time it reaches the edge of the grate and falls into the ash collection system. In the spreader stoker, where the fuel is thrown onto the grate, the smaller coal particles tend to fluidize and burn a short distance above the grate. Larger particles fall onto the grate and combust there.

Difficulties that must be overcome with stoker units include grate overheating and ash clinkering. Grate cooling may be accomplished in several ways. Air-cooled grates use the flow of underfire air to protect the equipment. The coal-to-ash bed that resides on the grate helps protect the metal from radiant heat above. Vibrating stoker grates may be water cooled, with cooling tubes connected to the waterwall network.

Ash that fuses together in large chunks is known by the term "clinkers." Clinkers can disrupt air patterns in the stoker and introduce abnormally large chunks of solidified material to the ash handling system. Uniform fuel distribution is a key in preventing clinker formation.

Stokers are not popular for new utility coal units, as the more modern techniques of pulverized coal firing and fluidized-bed combustion offer better efficiency. However, stokers are one of the viable methods to combust fuels that are difficult or impossible to reduce in size. These include biomass and refuse-derived fuel (RDF).

Biomass includes wood, bark, agricultural by-products such as corn stalks, and even more exotic materials like bamboo stalks and rice hulls. A very interesting fact about biomass is that growing plants absorb almost as much carbon dioxide as they give off during combustion. This offers potential in the fight to stabilize $CO_2$ emissions.

RDF-powered generation and co-generation offer methods to dispose of waste other than by landfilling. RDF is a difficult material to handle and burn. Fuel quality is quite variable and combustion produces significant chloride concentrations in the furnace, which can cause serious corrosion in the boiler and boiler backpass. Techniques to combat corrosion include the use of refractory in high heat areas of the furnace and more exotic alloys for boiler and superheater tube material.

A new process for refuse mass burning that combines aspects of fluidized bed combustion and stokers is shown in Figure 2-12. This is the Aireal™ process patented by Barlow Projects, Inc. Refuse is conveyed to a sharply inclined grate that has no moving parts like conventional stokers. Gravity does much of the work in moving the material. Underfire air flows upwards through the grate, but a unique feature is a series of air discharge ports along the length of the grate, from which sequenced pulses of air augment gravity to keep fuel flowing as it travels from top to bottom. The periodic pulses improve fuel/air mixing and combustion, but are regulated to prevent fuel from being prematurely blown to the bottom of the grate. The pulsing system adds an additional fuel flow control method previously not available for solid fuel combustion. The hot combustion gases flow to a heat recovery steam generator for steam production that may be used to drive a turbine for power generation. Backend equipment on the unit includes a lime-fed acid gas scrubber and fabric filters for particulate and heavy metals removal. The

**Fig. 2-12:** Illustration of the Aireal™ process, patented by Barlow Projects, Inc.

use of fabric filters allows for activated carbon injection into the flue gas stream to absorb metals.

## Coal gasification

Integrated coal gasification/power production is an emerging technology. It traces back to German scientists who developed coal gasification for production of synthetic fuels in the 1930s. Later in the century, Dow Chemical Company took the process further by using coal gasification to produce power at the Plaquemine facility in Louisiana. Modern coal gasification/power production projects—many of which are derivatives of the Dow effort—have largely been funded by the U.S. Department of Energy (DOE) as part of their Clean Coal Technology program.

The chemistry behind coal gasification is rather complex, but when viewed in simplistic terms, is readily understandable. First, consider combustion of coal or any other fossil fuel in a conventional boiler. When the fuel is burned with sufficient oxygen, it oxidizes completely as follows:

$$C + O_2 \rightarrow CO_2$$

Likewise, hydrogen in the coal combusts to water and sulfur burns to sulfur dioxide, which of course is a pollutant when released to the atmosphere. If, however, coal is heated in a moisture-laden (water or steam) environment containing a sub-stoichiometric ratio of oxygen, the product evolves into a mixture of carbon dioxide, carbon monoxide ($CO$), hydrogen ($H_2$), and smaller quantities of hydro-

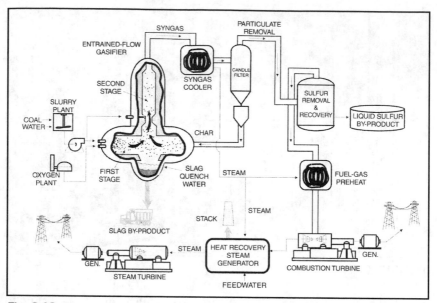

**Fig. 2-13:** Schematic of an integrated coal gasification combined-cycle process (Source: *Clean Coal Technology Program: Program Update 2000*, published by the U.S. Department of Energy)

gen sulfide ($H_2S$), carbonyl sulfide (COS), and ammonia ($NH_3$). This gas has a significant heating value that may be productively used elsewhere.

A good example of how coal gasification is utilized to produce power is illustrated by the DOE integrated coal gasification combined-cycle (ICGCC) repowering project at the Wabash River Plant in Indiana. The system was set up to repower an existing steam turbine and to add new generating capacity. Figure 2-13 outlines the initial design. The most important steps include the following:

- A pulverized-coal/water slurry is fed to the lower section of a refractory-lined gasifier. Partial combustion of the coal takes place in this zone. The combustion temperature of 2,500°F (1,371°C) exceeds the ash melting point, so slag flows to the bottom of the unit for extraction

- At a point higher in the combustor, a second slurried coal feed enters. As it mixes with the hot gas rising from below, the fresh coal devolatilizes and breaks down, producing the final synthetic gas mixture

- The syngas passes through a heat exchanger, which not only cools the gas but produces steam at 1,600 psia (11.03 mPa) for power production. The gas then passes through filters that remove particulates, including unburned carbon. These are returned to the gasifier

- The syngas passes through a catalytic converter to convert COS to $H_2S$ and is then scrubbed with an alkaline amine solution to remove the $H_2S$. The $H_2S$-rich amine solution is routed through a steam stripper that removes the $H_2S$. The regenerated amine returns to the process, while the $H_2S$ is sent to a conventional Claus unit for conversion to elemental sulfur

- The clean syngas passes through a fuel gas preheater, and then is injected into a gas turbine for power production

- In the classical combined-cycle arrangement, waste heat from the gas turbine generates steam in a single-pressure HRSG for additional power production

Some basic data, as reported by the DOE, are illustrated in Tables 2-1 and 2-2. Of note is the thermal efficiency of almost 40%, which favorably compares with

| | |
|---|---|
| Gas Turbine Power Ouput (Mwe) | 192 |
| Repowered Steam Turbine Power Output (Mwe) | 104 |
| Parasitic Power Loss (Mwe) | 34 |
| Main Steam Pressure (psig) | 1585 |
| Combustor Pressure (psig) | 400 |
| Maximum Superheater Steam Flow (lb/hr) | 754,000 |
| Superheater Steam Temperature (°F) | 1010 |
| Maximum Reheater Steam Flow (lb/hr) | 600,820 |
| Reheater Steam Temperature (°F) | 1010 |
| Unit Efficiency (%) | 39.7 |

**Table 2-1**: IGCC Plant data

| | $SO_2$ | $NO_x$ | CO | PM-10 | VOC |
|---|---|---|---|---|---|
| Pre-IGCC Emissions Rates | 3.1 | 0.8 | 0.05 | 0.07 | 0.003 |
| IGCC Emissions Rates | 0.1 | 0.15 | 0.05 | ND | 0.003 |

**Table 2-2**: Emissions from original boiler and IGCC unit at Wabash RIver (values reported in Ib/MBtu)

conventional boilers. Also of great importance are the emissions reductions, particularly for $SO_2$ and $NO_x$. Both are well below current and projected future requirements.

The Wabash River unit started up in 1995 and received final approval in 2000. The DOE reported on a number of problems that had to be solved during the commissioning period. First, the gasification process generated vaporous chlorides that poisoned the COS-$H_2S$ conversion catalyst. Retrofit of an additional scrubbing system removed the chlorides ahead of the catalyst bed. Candle filters for particulate removal were originally ceramic, but these fractured due to stresses in the system. Metallic filters solved this problem. Ash deposits in the fire-tubed flue gas cooler created difficulties that were solved by modifications to the flow path and geometry, and by periodic mechanical cleaning of the tubes. The $H_2S$ removal system was initially undersized; capacity was increased. Finally, the maximum availability for the unit for any extended period of time has only been 77%—less than many conventional units. However, an advantage of the combined-cycle concept is the gas turbine may be fired with natural gas during periods when the gasifier and its auxiliaries are down for maintenance.

Perhaps the most important point about integrated coal gasification combined-cycle systems is that coal rather than natural gas is the primary fuel in a combined-cycle process. Energy personnel, politicians, and the media frequently comment on the need for a balanced energy policy. The ICGCC technique appears to offer good potential for the continued use of coal as a power generation fuel.

# Appendix 2-1

Repeated below are the two most important reactions regarding the $SO_2$ removal process in a circulating fluidized-bed boiler.

$$CaCO_3 + heat \rightarrow CaO + CO_2 \uparrow$$

$$CaO + SO_2 + \tfrac{1}{2} O_2 \rightarrow CaSO_4 + heat$$

When the University of North Carolina at Chapel Hill (UNC-CH) changed limestone in their twin CFBs, limestone consumption dropped drastically. At the same time, bed temperatures in the two units rose approximately 60°F to 70°F (16°C to 21°C), and $NO_x$ emission levels increased significantly, as well. Test personnel concluded that the bed temperatures rose due to the greater reactivity of the limestone.

Prior to the change, a significant overfeed of limestone was required to remove the required amount of $SO_2$. The extra calcium carbonate calcined per the first equation outlined above, but did not react with $SO_2$. The calcining process did, however, absorb heat. Once the change was made, limestone feed dropped almost by half, and the corresponding heat absorption also declined, resulting in increased bed temperatures. Utility personnel had to adjust fuel flow and other parameters to return the beds to normal temperatures.

# Appendix 2-2

Flow-assisted corrosion (FAC) is a phenomenon affecting many steam generating units, causing fatalities at several plants. FAC occurs at flow disturbances, such as low-pressure HRSG evaporator tubes, feedwater pipe elbows, and economizer elbows, where the feedwater or boiler water has been heavily dosed with an oxygen scavenger, and where the temperature is near 300°F (149°C). The flowing liquid gradually dissolves the protective magnetite layer at the point of attack, leading to wall thinning and eventual pressure-induced failure (Fig. A2-1).

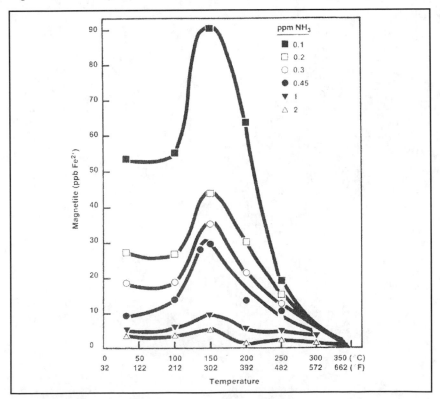

**Fig. A2-1:** Solubility of magnetite in ammonia (Source: EPRI)

Typical HRSG configurations require many short-radius boiler tube elbows. The low-pressure evaporator, where temperatures may be at or near 300°F, can be particularly susceptible to FAC. Current and future plant designers, engineers, managers, and owners need to be aware of this problem and act accordingly. The Electric Power Research Institute (EPRI; www.epri.com) has published a book on FAC, and FAC has also been a prominent discussion topic at the International Water Conference and the Electric Utility Chemistry Workshop.

# Chapter 3

## Fossil Fuel and Ash Properties–Their Effects on Steam Generator Materials

## INTRODUCTION

As was generally outlined in the first chapter, fuel type and ash properties have a great impact on boiler size, volume, tube arrangement, firing method, and other parameters. Fossil fuel composition and combustion products also influence selection of boiler materials. This chapter examines fuel and ash properties in more detail.

## COAL

Coal is compressed plant matter that over millions of years transformed into a high-carbon material. Age, type of initial vegetation, and location of deposit formation are all significant factors in the quality of a coal deposit. Scientific research indicates that the first plants to grow on land appeared more than 400 million years ago during the Silurian and Devonian periods of the Paleozoic Era. This early vegetation consisted mostly of leafless shoots. Some 375 million years ago, extensive forests covered much of the world, and it is approximately from this time that coal deposits have been dated. The period from 350 to 275 million years ago is known as the Carboniferous period, and during this time many coal deposits originated. This was a period of globally warm temperatures, which encouraged plant growth. Indeed, plants and vegetation grew to unimaginable sizes when compared to our current world. Giant ferns and plants the size of current-day mature trees were quite common.

At the end of the Carboniferous period—and for about 135 million years thereafter—coal formation in the Northern Hemisphere greatly diminished. With the onset of the Cretaceous period in the Mesozoic Era (around 135 million B.C.), plant growth and coal formation resumed, although by this time more complex vegetation, including plants and trees with protected seeds, had begun to dominate the landscape. Thus, the coal that we use today developed over hundreds of millions of years from a wide variety of vegetation. What was the process behind coal formation?

To understand the general properties of coal, it is first necessary to understand the basic chemical composition of plant life. What is important to know is that the main building block of vegetation is cellulose. Cellulose belongs to a class of compounds known as carbohydrates, whose name comes from the fact that the compounds are composed of carbon, oxygen, and hydrogen. Cellulose fibers within a plant are held in place and bonded by another carbon-based polymer known as lignin. Together, cellulose and lignin comprise the bulk of plant material, although other natural compounds such as hemicellulose and resin are present. In each compound, the primary elements are carbon, oxygen, and hydrogen.

The prerequisite for coal formation were the vast swamps that covered much of the earth in prehistoric times. Vegetation that dies upon firm ground is decomposed by the atmosphere into carbon dioxide and water. When vegetation dies in a swamp, a much different process occurs. The first step is bacterial attack of the dead vegetation. Microorganisms consume hydrogen and oxygen, increasing the carbon content. This mechanism, known as the biochemical phase of coalification, is self-limiting, as the bacterial action produces organic compounds eventually becoming lethal to the organisms themselves. Over time, the partially decomposed matter becomes overlaid by other material, including more vegetation and soil. This process has two principal effects—it places the material under increasing pressure and moves the deposits deeper underground where temperatures are warmer.

| Material | Composition by %wt. (DAF Basis) | | |
|---|---|---|---|
| | Carbon | Hydrogen | Oxygen |
| Wood | 49 | 7 | 44 |
| Peat | 60 | 6 | 34 |
| Lignite | 70 | 5 | 25 |
| Subbituminous | 75 | 5 | 20 |
| Bituminous | 85 | 5 | 10 |
| Anthracite | 94 | 3 | 3 |

**Table 3-1:** Change in chemical composition as a result of coalification (Reproduced with permission from *Steam*, 40th ed., published by Babcock & Wilcox, a McDermott Company)

The combination of pressure and heat causes additional loss of oxygen and hydrogen. This is known as the geochemical phase of coalification. The results are graphically illustrated in Table 3-1, which shows the primary chemical composition of the plant material from wood to the most mature of coals—anthracite. The principal point is that as a coal matures, carbon content increases. Theoretically, a completely mature coal would have the chemical composition of graphite.

Complex carbohydrates within vegetation are created from sugars and starches that the plant produces through photosynthesis. During the coalification process, these compounds and others metamorphose to low-weight organic molecules that are not bound to the main coal structure. The smaller organic compounds are known as volatiles because they vaporize with increasing temperature. Volatiles are driven off during the coalification process, and increasingly mature coals contain less volatile content.

One might be tempted to think that age is a primary factor in the maturity of coal. While this is true in some cases, the two most important factors are pressure and temperature. Coals that were buried deep and located in high temperature zones—underneath a region of volcanic activity—mature much more quickly than older coals subjected to less heat and pressure.

| Class | Group | Fixed Carbon Limits, % (Dry, Mineral-Matter-Free Basis) Equal or Greater Than | Less Than | Volatile Matter Limits, % (Dry, Mineral-Matter-Free Basis) Greater Than | Equal or Less Than | Calorific Value Limits, Btu/lb (Moist,[b] Mineral-Matter-Free Basis) Equal or Greater Than | Less Than | Agglomerating Character |
|---|---|---|---|---|---|---|---|---|
| I. Anthracitic | 1. Meta-anthracite | 98 | — | — | 2 | — | — | Nonagglomerating |
| | 2. Anthracite | 92 | 98 | 2 | 8 | — | — | |
| | 3. Semianthracite[c] | 86 | 92 | 8 | 14 | — | — | |
| II. Bituminous | 1. Low volatile bituminous coal | 78 | 86 | 14 | 22 | — | — | Commonly agglomerating[e] |
| | 2. Medium volatile bituminous coal | 69 | 78 | 22 | 31 | — | — | |
| | 3. High volatile A bituminous coal | — | 69 | 31 | — | 14,000[d] | — | |
| | 4. High volatile B bituminous coal | — | — | — | — | 13,000[d] | 14,000 | |
| | 5. High volatile C bituminous coal | — | — | — | — | 11,500 | 13,000 | |
| | | | | | | 10,500[e] | 11,500 | Agglomerating |
| III. Subbituminous | 1. Subbituminous A coal | — | — | — | — | 10,500 | 11,500 | Nonagglomerating |
| | 2. Subbituminous B coal | — | — | — | — | 9,500 | 10,500 | |
| | 3. Subbituminous C coal | — | — | — | — | 8,300 | 9,500 | |
| IV. Lignitic | 1. Lignite A | — | — | — | — | 6,300 | 8,300 | |
| | 2. Lignite B | — | — | — | — | — | 6,300 | |

[a]This classification does not include a few coals, principally nonbanded varieties, which have unusual physical and chemical properties and which come within the limits of fixed carbon or calorific value of the high volatile bituminous and subbituminous ranks. All of these coals either contain less than 48% dry, mineral-matter-free Btu/lb.

[b]Moist refers to coal containing its natural inherent moisture but not including visible water on the surface of the coal.

[c]If agglomerating, classify in low volatile group of the bituminous class.

[d]Coals having 69% or more fixed carbon on the dry, mineral-matter-free basis shall be classified according to fixed carbon, regardless of calorific value.

[e]It is recognized that there may be nonagglomerating varieties in these groups of the bituminous class, and there are notable exceptions in high volatile C bituminous group.

**Table 3-2:** Classification of coals by rank ASTM D 388 (Reproduced with permission from *Steam*, 40th ed., published by Babcock & Wilcox, a McDermott Company)

| No. | Coal Rank Class | Coal Rank Group | State | County | Coal Analysis, Bed Moisture Basis M | VM | FC | A | S | Btu | Rank FC | Rank Btu |
|---|---|---|---|---|---|---|---|---|---|---|---|---|
| 1 | I | 1 | Pa. | Schuylkill | 4.5 | 1.7 | 84.1 | 9.7 | 0.77 | 12,745 | 99.2 | 14,280 |
| 2 | I | 2 | Pa. | Lackawanna | 2.5 | 6.2 | 79.4 | 11.9 | 0.60 | 12,925 | 94.1 | 14,880 |
| 3 | I | 3 | Va. | Montgomery | 2.0 | 10.6 | 67.2 | 20.2 | 0.62 | 11,925 | 88.7 | 15,340 |
| 4 | II | 1 | W.Va. | McDowell | 1.0 | 16.6 | 77.3 | 5.1 | 0.74 | 14,715 | 82.8 | 15,600 |
| 5 | II | 1 | Pa. | Cambria | 1.3 | 17.5 | 70.9 | 10.3 | 1.68 | 13,800 | 81.3 | 15,595 |
| 6 | II | 2 | Pa. | Somerset | 1.5 | 20.8 | 67.5 | 10.2 | 1.68 | 13,720 | 77.5 | 15,485 |
| 7 | II | 2 | Pa. | Indiana | 1.5 | 23.4 | 64.9 | 10.2 | 2.20 | 13,800 | 74.5 | 15,580 |
| 8 | II | 3 | Pa. | Westmoreland | 1.5 | 30.7 | 56.6 | 11.2 | 1.82 | 13,325 | 65.8 | 15,230 |
| 9 | II | 3 | Ky. | Pike | 2.5 | 36.7 | 57.5 | 3.3 | 0.70 | 14,480 | 61.3 | 15,040 |
| 10 | II | 3 | Ohio | Belmont | 3.6 | 40.0 | 47.3 | 9.1 | 4.00 | 12,850 | 55.4 | 14,380 |
| 11 | II | 4 | Ill. | Williamson | 5.8 | 36.2 | 46.3 | 11.7 | 2.70 | 11,910 | 57.3 | 13,710 |
| 12 | II | 4 | Utah | Emery | 5.2 | 38.2 | 50.2 | 6.4 | 0.90 | 12,600 | 57.3 | 13,560 |
| 13 | II | 5 | Ill. | Vermilion | 12.2 | 38.8 | 40.0 | 9.0 | 3.20 | 11,340 | 51.8 | 12,630 |
| 14 | III | 1 | Mont. | Musselshell | 14.1 | 32.2 | 46.7 | 7.0 | 0.43 | 11,140 | 59.0 | 12,075 |
| 15 | III | 2 | Wyo. | Sheridan | 25.0 | 30.5 | 40.8 | 3.7 | 0.30 | 9,345 | 57.5 | 9,745 |
| 16 | III | 3 | Wyo. | Campbell | 31.0 | 31.4 | 32.8 | 4.8 | 0.55 | 8,320 | 51.5 | 8,790 |
| 17 | IV | 1 | N.D. | Mercer | 37.0 | 26.6 | 32.2 | 4.2 | 0.40 | 7,255 | 55.2 | 7,610 |

Notes: For definition of Rank Classification according to ASTM requirements, see Table 3.

Data on Coal (Bed Moisture Basis)

M = equilibrium moisture, %; VM = volatile matter, %;
FC = fixed carbon, %; A = ash, %; S = sulfur, %;
Btu = Btu/lb, high heating value.

Rank FC = dry, mineral-matter-free fixed carbon, %;
Rank Btu = moist, mineral-matter-free Btu/lb.
Calculations by Parr formulas.

**Table 3-3:** Properties of some U.S. coals (Reproduced with permission from *Steam*, 40th ed., published by Babcock & Wilcox, a McDermott Company)

Tables 3-2 and 3-3 list the ASTM classification and characteristics of coals. Let us use them to examine fundamental chemical properties and heating values of coal. We will then take a close look at the impurities that reside within coal deposits, and how they behave during the combustion process.

The natural maturation process during coalification is:

wood → peat → lignite → subbituminous → bituminous → anthracite

Examples of the different types of coal can be found throughout the world. In the U.S., the Appalachian area around western Pennsylvania, West Virginia, Ohio, Kentucky, and stretching into Alabama contains enormous deposits of bituminous coal. The state of Illinois also sits atop an extensive bituminous deposit. A large subbituminous deposit resides beneath the states of Wyoming and Montana, and because much of this coal is mined in an area near the Powder River, is known as Powder River Basin (PRB) coal. Significant lignite deposits are located in North Dakota and to a lesser extent in Texas. Around the world, China has large bituminous and lignite deposits; Russia has enormous deposits of bituminous coal and some lignite in many different areas; Germany has significant deposits of bituminous and brown coal (an immature lignite); and Great Britain is endowed with large reserves of bituminous and anthracite. In the southern hemisphere, Australia has significant deposits of bituminous and brown coal.

# Peat

A visual examination of peat provides a clear example of the intermediate stage between plant life and coal deposits. Peat may range from a light-colored substance that has recognizable pieces of plant matter to a black material that looks like coal. Although peat is continually being compacted by overlying material, it is still subject to microbiological decomposition in the biochemical phase of coalification. One of the byproducts of this process is methane, which is commonly referred to as "swamp gas" or "marsh gas."

The typical aging process for peat involves a general rule that it takes 100 years for a 2- to 3-inch layer of peat to form. Some modern swamps have peat layers up to 30 feet deep, which means they have been undisturbed for thousands of years. Not uncommon are layered coal deposits, where each seam is separated by soil and minerals. This suggests that some ancient swamps produced a layer of peat, died out, then redeveloped to start the process over again.

# Lignite

Continued compression and heating of peat produce lignite and its more immature precursor—brown coal. While peat is not considered to be coal, lignite definitely falls into the coal category, although plant material is often still clearly evident in lignite deposits. Carbon content in lignite is around 70%, while oxygen content has dropped to 25%. This is the first fuel listed in the ASTM coal classification table, and as is clearly evident, the heating value (quantity of energy available from combustion) is the lowest of all coals, with a range of 6,300 to 8,300 Btu per pound (14,653–19,305 kJ per kilogram) on a moisture and ash free basis.

Lignite-fired boilers are typically much larger in size than other boilers because long residence times are required to extract the energy from the fuel. Lignite is not a common fuel of choice. It is mostly used at mine-mouth power plants, in which the fuel is conveyed directly from the mine to the plant. Lignites contain much volatile matter and are the easiest coals to ignite. For the same reason, they are also the coal most prone to spontaneous combustion in coal piles and bunkers (Appendix 3-1). Although lignite has much less moisture than peat (30% as compared to 70%), the water content is still quite high. This must be taken into account when designing fuel handling and drying systems for lignite.

## Subbituminous

The next step in the evolutionary stage of coal formation is subbituminous. This coal has a 75% carbon content with only 20% oxygen. Like lignite, the ASTM ranks this coal on energy content. Subbituminous coal has a high volatile content and is subject to spontaneous combustion when stored improperly or too long in piles or coal bunkers. The high volatile content gives subbituminous coal good ignition properties within a boiler.

## Bituminous

Bituminous coal offers advantages of high heating content and enough volatiles to ignite quickly in a boiler, with the propensity to severe cases of spontaneous combustion. Bituminous coal also has just the right properties for coke production, which is vital to the steel industry.

Other factors aside, bituminous coal would be the preferred fuel for many coal-fired boilers, except for the presence of significant sulfur concentrations in many bituminous deposits making it less attractive.

## Anthracite

Anthracite represents the most mature coal. Volatile matter is very slight, making the coal difficult to ignite. The only significant deposit of anthracite in the U.S. is located in Pennsylvania, so other than home heating use in the nineteenth and part of the twentieth centuries, anthracite has not been heavily exploited as an energy source in this country.

## Impurities in coal

Some impurities accumulated in coal during original plant growth, but most came from external sources that were strongly dependent upon where or in what conditions the coal formed.

Two natural elements in coal are sulfur and nitrogen, both of which were contained in the amino acids and metabolites produced by the original vegetation. Fuel-bound sulfur generally accounts for only a small portion of the sulfur in a coal deposit, as most of it exists in mineral form. Fuel-bound nitrogen is the main contributor in $NO_x$ formation during combustion, which will be examined more closely in chapter 5.

Water infiltration through coal seams and crevices is in fact the primary source of coal impurities. Peat, of course, has very high moisture content, but even mature coals, including bituminous, contain many cracks and crevices allowing the passage of water. The minerals most commonly found in coals are listed in Table 3-4. A very common impurity is iron sulfide ($FeS_2$). Iron sulfide is believed to have come from swamps originally flooded with brackish water containing sulfates; anaerobic bacterial breakdown of the sulfates produced sulfides, which combined with iron. $FeS_2$ is usually responsible for most of the sulfur within coal and is the most troublesome impurity of many eastern bituminous coals.

Soil and many natural minerals consist of complex metallic silicates, so virtually all coals contain silicon and aluminum in significant quantities. Calcium and magnesium may reach relatively high proportions if the coal is located near limestone deposits. Sodium and potassium are other important impurities, as they can cause a great deal of trouble in the backpasses of a boiler.

| | |
|---|---|
| Kaolinite | $Al_2O_3 \cdot 2SiO_2 \cdot H_2O$ |
| Illite | $K_2O \cdot 3Al_2O_3 \cdot 6SiO_2 \cdot 2H_2O$ |
| Muscovite | $K_2O \cdot 3Al_2O_3 \cdot 6SiO_2 \cdot 2H_2O$ |
| Biotite | $K_2O \cdot MgO \cdot Al_2O_3 \cdot 3SiO_2 \cdot H_2O$ |
| Orthoclase | $K_2O \cdot Al_2O_3 \cdot 6SiO_2$ |
| Albite | $Na_2O \cdot Al_2O_3 \cdot 6SiO_2$ |
| Calcite | $CaCO_3$ |
| Dolomite | $CaCO_3 \cdot MgCO_3$ |
| Siderite | $FeCO_3$ |
| Pyrite | $FeS_2$ |
| Gypsum | $CaSO_4 \cdot 2H_2O$ |
| Quartz | $SiO_2$ |
| Hematite | $Fe_2O_3$ |
| Magnetite | $Fe_3O_4$ |
| Rutile | $TiO_2$ |
| Halite | $NaCl$ |
| Sylvite | $KCl$ |

**Table 3-4:** Common minerals found in coal (Source: Sinder, J.G., ed., *Combustion: Fossil Power*, Alstom)

## Analyzing coal

The two common analytical methods for determining the makeup of coal are the proximate analysis and the ultimate analysis. The ASTM has published specific guidelines for both in ASTM D3172 and D3176, respectively.

# Basics of Boiler & HRSG Design

| | Anthracite | Pittsburgh #8 HV Bituminous | Illinois #6 HV Bituminous | Upper Freeport MV Bituminous | Spring Creek Subbituminous | Decker Subbituminous | Lignite | Lignite (S. Halleville) | Lignite (Bryan) | Lignite (San Miguel) |
|---|---|---|---|---|---|---|---|---|---|---|
| State | — | Ohio or Pa. | Illinois | Pennsylvania | Wyoming | Montana | North Dakota | Texas | Texas | Texas |
| **Proximate:** | | | | | | | | | | |
| Moisture | 7.7 | 5.2 | 17.6 | 2.2 | 24.1 | 23.4 | 33.3 | 37.7 | 34.1 | 14.2 |
| Volatile matter, dry | 6.4 | 40.2 | 44.2 | 28.1 | 43.1 | 40.8 | 43.6 | 45.2 | 31.5 | 21.2 |
| Fixed carbon, dry | 83.1 | 50.7 | 45.0 | 58.5 | 51.2 | 54.0 | 45.3 | 44.4 | 18.1 | 10.0 |
| Ash, dry | 10.5 | 9.1 | 10.8 | 13.4 | 5.7 | 5.2 | 11.1 | 10.4 | 50.4 | 68.8 |
| **Heating value, Btu/lb:** | | | | | | | | | | |
| As-received | 11,890 | 12,540 | 10,900 | 12,970 | 9,190 | 9,540 | 7,090 | 7,080 | 3,930 | 2,740 |
| Dry | 12,880 | 13,230 | 12,500 | 13,260 | 12,110 | 12,456 | 10,630 | 11,360 | 5,960 | 3,200 |
| MAF | 14,390 | 14,550 | 14,010 | 15,330 | 12,840 | 13,130 | 11,960 | 12,680 | 12,020 | 10,260 |
| **Ultimate:** | | | | | | | | | | |
| Carbon | 83.7 | 74.0 | 69.0 | 74.9 | 70.3 | 72.0 | 63.3 | 66.3 | 36.8 | 18.4 |
| Hydrogen | 1.9 | 5.1 | 4.9 | 4.7 | 5.0 | 5.0 | 4.5 | 4.9 | 3.3 | 2.3 |
| Nitrogen | 0.9 | 1.6 | 1.0 | 1.27 | 0.96 | 0.95 | 1.0 | 1.0 | 0.4 | 0.29 |
| Sulfur | 0.7 | 2.3 | 4.3 | 0.76 | 0.35 | 0.44 | 1.1 | 1.2 | 1.0 | 1.2 |
| Ash | 10.5 | 9.1 | 10.8 | 13.4 | 5.7 | 5.2 | 11.1 | 10.4 | 50.4 | 68.8 |
| Oxygen | 2.3 | 7.9 | 10.0 | 4.97 | 17.69 | 16.41 | 19.0 | 16.2 | 11.1 | 9.01 |

**Ash fusion temps, F** (Reducing/Oxidizing: Red / Oxid)

| | Anthr. R | Anthr. O | Pitt. R | Pitt. O | Ill. R | Ill. O | U.Free. R | U.Free. O | S.Creek R | S.Creek O | Decker R | Decker O | Lignite R | Lignite O | S.Hall. R | S.Hall. O | Bryan R | Bryan O | S.Mig. R | S.Mig. O |
|---|---|---|---|---|---|---|---|---|---|---|---|---|---|---|---|---|---|---|---|---|
| ID | — | — | 2220 | 2560 | 1930 | 2140 | 2750+ | 2750+ | 2100 | 2180 | 2120 | 2420 | 2030 | 2160 | 2000 | 2210 | 2370 | 2470 | 2730 | 2750+ |
| ST Sp. | — | — | 2440 | 2640 | 2040 | 2330 | " | " | 2160 | 2300 | 2250 | 2470 | 2130 | 2190 | 2060 | 2250 | 2580 | 2670 | 2750+ | " |
| ST Hsp. | — | — | 2470 | 2650 | 2080 | 2400 | " | " | 2170 | 2320 | 2270 | 2490 | 2170 | 2220 | 2090 | 2280 | 2690 | 2760 | " | " |
| FT 0.0625 in. | — | — | 2570 | 2670 | 2420 | 2600 | " | " | 2190 | 2360 | 2310 | 2510 | 2210 | 2280 | 2220 | 2350 | 2900+ | 2900+ | " | " |
| FT Flat | — | — | 2750+ | 2750+ | 2490 | 2700 | " | " | 2370 | 2700 | 2380 | 2750+ | 2300 | 2300 | 2330 | 2400 | 2900+ | 2900+ | " | " |

| | Anthracite | Pittsburgh #8 HV Bituminous | Illinois #6 HV Bituminous | Upper Freeport MV Bituminous | Spring Creek Subbituminous | Decker Subbituminous | Lignite | Lignite (S. Halleville) | Lignite (Bryan) | Lignite (San Miguel) |
|---|---|---|---|---|---|---|---|---|---|---|
| **Ash analysis:** | | | | | | | | | | |
| $SiO_2$ | 51.0 | 50.58 | 41.68 | 59.60 | 32.61 | 23.77 | 29.80 | 23.22 | 62.4 | 66.85 |
| $Al_2O_3$ | 34.0 | 24.62 | 20.0 | 27.42 | 13.38 | 15.79 | 10.0 | 13.0 | 21.5 | 23.62 |
| $Fe_2O_3$ | 3.5 | 17.16 | 19.0 | 4.67 | 7.53 | 6.41 | 9.0 | 22.0 | 3.0 | 1.18 |
| $TiO_2$ | 2.4 | 1.10 | 0.8 | 1.34 | 1.57 | 1.08 | 0.4 | 0.8 | 0.5 | 1.46 |
| $CaO$ | 0.6 | 1.13 | 8.0 | 0.62 | 15.12 | 21.85 | 19.0 | 22.0 | 3.0 | 1.76 |
| $MgO$ | 0.3 | 0.62 | 0.8 | 0.75 | 4.26 | 3.11 | 5.0 | 5.0 | 1.2 | 0.42 |
| $Na_2O$ | 0.74 | 0.39 | 1.62 | 0.42 | 7.41 | 6.20 | 5.80 | 1.05 | 0.59 | 1.67 |
| $K_2O$ | 2.65 | 1.99 | 1.63 | 2.47 | 0.87 | 0.57 | 0.49 | 0.27 | 0.92 | 1.57 |
| $P_2O_5$ | — | 0.39 | — | 0.42 | 0.44 | 0.99 | | | | |
| $SO_3$ | 1.38 | 1.11 | 4.41 | 0.99 | 14.56 | 18.85 | 20.85 | 9.08 | 3.50 | 1.32 |

Note: HV = high volatile; MV = medium volatile; ID = initial deformation temp; ST = softening temp; FT = fluid temp; Sp. = spherical; Hsp. = hemispherical.

**Table 3-5:** Properties of U.S. coals including ash analyses (Reproduced with permission from *Steam*, 40th ed., published by Babcock & Wilcox, a McDermott Company)

The proximate analysis determines moisture, volatile matter, fixed carbon, ash, and the heating value. The ultimate analysis determines the basic chemical content of coal and produces values for carbon, hydrogen, nitrogen, sulfur, ash, and oxygen. Table 3-5 illustrates the proximate and ultimate analyses for several U.S. coals. The table also includes ash analyses and ash fusion temperatures, which will be examined in a later section.

Moisture in coal can be either free or combined. Free moisture is not chemically bound to the coal and evaporates fairly readily when the coal is exposed to air. Combined moisture is that which exists within molecular structures. Mineral compounds in coal may retain some of this moisture. A well-known example of a chemical that contains combined moisture is gypsum, the material used in wallboard. Gypsum is also a common impurity in coal. The formula for gypsum is $CaSO_4 \cdot 2H_2O$, in which the water molecules are trapped within the crystal lattice of the calcium sulfate. This moisture does not escape at ambient temperatures, but is released only when the molecule is heated to 262°F (128°C) and beyond.

Bomb calorimetry is used for determining the heating value. This method gives the higher heating value (HHV) of the sample, which is the heat required to convert the entire coal sample to a gaseous state. However, the vaporization of the inherent moisture in the coal and moisture formed during the combustion process do not provide heating value, so the energy consumed by these processes is deducted from the HHV to give the lower heating value (LHV).

Heating the sample to different temperatures and then weighing the residual matter determines volatile matter, fixed carbon, and ash. Heating in the absence of air drives off the volatile matter but leaves fixed carbon and ash. Subsequent heating with air burns the fixed carbon, leaving only the ash. Many power plant laboratories have the capability to analyze sulfur and include the results with proximate analyses. A principal criterion for monitoring sulfur dioxide emission in the U.S. is "pounds of $SO_2$ per million Btu of heat input," and this is a value a lab needs to determine for most, if not all proximate analyses.

The ultimate analysis provides a more detailed breakdown of the primary coal components, although it does not separately list the minerals in ash. Outside laboratories usually analyze samples for carbon, hydrogen, nitrogen, and oxygen.

# OIL

While coal is the biologically/chemically altered remains of plant life, oil is the remnant of marine animals. In another process that took millions of years to complete, sea creatures buried with mud and silt decomposed under temperature and pressure to produce organic deposits. In this case, the composition of the sea creatures differed from that of vegetation; and the breakdown produced smaller organic molecules that liquefied. Often, the liquefied molecules filtered through the rock formations to collect in pockets. These pockets are the sources of the "gushers" that American prospectors once found in the U.S., and which can still be found in other areas of the world. In other cases, the oil became trapped within the sediment and is not readily extractable. A classic example is a geologist setting a piece of shale on fire due to the minute droplets of oil contained within the rock.

Oil, more than any other natural resource, is a driving force in international politics. For the U.S. power industry, however, oil-fired generation accounts for a very small percentage of total energy production—less than 3%. It once was higher, but the decline in U.S. oil reserves coupled with the volatility of supplies from the Middle East forced utilities to switch to alternate energy sources.

The number and type of petrochemicals that we have developed from oil are enormous. The classic examples are plastics. With regard to fuels, oil contains a wide number of various sized hydrocarbons, many of which can be separated by distillation. Through this process and those of catalytic cracking, catalytic reforming, and others, fuels can be produced ranging from the very light hydrocarbons to heavy fuel oils. Tables 3-6 and 3-7 outline the properties of the primary fuel oils. These tables illustrate such items as flash point, pour point, viscosity, and general chemical makeup. Definitions for some of these properties are listed in Table 3-8.

The most common use for oil in power generation today is as a "light-off fuel" to start coal-fired plants. Number 2 fuel oil fits the bill well, as it is a fairly free-flowing fluid. The fuel oils have a significantly higher heating value per unit weight than coal. The fluid also serves as a fuel for combustion turbines, most often as a backup source to natural gas.

No. 1 A distillate oil intended for vaporizing pot-type burners and other burners requiring this grade of fuel

No. 2 A distillate oil for general purpose domestic heating for use in burners not requiring No. 1 fuel oil

No. 4 Preheating not usually required for handling or burning

No. 5 (Light) Preheating may be required depending on climate and equipment

No. 5 (Heavy) Preheating may be required for burning and, in cold climates, may be required for handling

No. 6 Preheating required for burning and handling

| Grade of Fuel Oil[b] | Flash Point, F (C) | Pour Point, F (C) | Water and Sediment, % by vol | Carbon Residue on 10% Bottoms, % | Ash, % by wt | Distillation Temperatures, F (C) | | | Saybolt Viscosity, s | | | | Kinematic Viscosity, centistokes | | | | Gravity, deg API | Copper Strip Corrosion |
|---|---|---|---|---|---|---|---|---|---|---|---|---|---|---|---|---|---|---|
| | | | | | | 10% Point | 90% Point | | Universal at 100F (38C) | | Furol at 122F (50C) | | At 100F (38C) | | At 122F (50C) | | | |
| | Min | Max | Max | Max | Max | Max | Min | Max | Min | Max | Min | Max | Min | Max | Min | Max | Min | Max |
| No. 1 | 100 or legal (38) | 0 | trace | 0.15 | — | 420 (215) | — | 550 (288) | — | … | — | — | 1.4 | 2.2 | — | — | 35 | No. 3 |
| No. 2 | 100 or legal (38) | 20[c] (−7) | 0.10 | 0.35 | — | [d] | 540[c] (282) | 640 (338) | (32.6)[f] | (37.93) | … | — | 2.0[e] | 3.6 | — | — | 30 | — |
| No. 4 | 130 or legal (55) | 20 (−7) | 0.50 | — | 0.10 | — | — | — | 45 | 125 | … | — | (5.8) | (26.4) | — | — | — | — |
| No. 5 (Light) | 130 or legal (55) | — | 1.00 | — | 0.10 | — | — | — | 150 | 300 | … | — | (32) | (65) | — | — | — | — |
| No. 5 (Heavy) | 130 or legal (55) | — | 1.00 | — | 0.10 | … | — | — | 350 | 750 | (23) | (40) | (75) | (162) | (42) | (81) | — | — |
| No. 6 | 150 (65) | — | 2.00[g] | — | — | — | — | — | (900) | (9,000) | 45 | 300 | — | — | (92) | (638) | — | — |

Notes:

a. Recognizing the necessity for low sulfur fuel oils used in connection with heat treatment, nonferrous metal, glass, and ceramic furnaces and other special uses, a sulfur requirement may be specified in accordance with the following table:

| Grade of Fuel Oil | Sulfur, Max, % |
|---|---|
| No. 1 | 0.5 |
| No. 2 | 0.7 |
| No. 4 | no limit |
| No. 5 | no limit |
| No. 6 | no limit |

Other sulfur limits may be specified only by mutual agreement between the purchaser and the seller.

b. It is the intent of these classifications that failure to meet any requirement of a given grade does not automatically place an oil in the next lower grade unless in fact it meets all requirements of the lower grade.

c. Lower or higher pour points may be specified whenever required by conditions of storage or use.

d. The 10% distillation temperature point may be specified at 440F (226C) maximum for use in other than atomizing burners.

e. When pour point less than 0F is specified, the minimum viscosity shall be 1.8 cs (32.0 s, Saybolt Universal) and the minimum 90% point shall be waived.

f. Viscosity values in parentheses are for information only and not necessarily limiting.

g. The amount of water by distillation plus the sediment by extraction shall not exceed 2.00%. The amount of sediment by extraction shall not exceed 0.50%. A deduction in quantity shall be made for all water and sediment in excess of 1.0%.

Source, ASTM D 396.

**Table 3-6:** Specifications for fuel oils (Reproduced with permission from *Steam*, 40th ed., published by Babcock & Wilcox, a McDermott Company)

| Grade of Fuel Oil | No. 1 | No. 2 | No. 4 | No. 5 | No. 6 |
|---|---|---|---|---|---|
| **% by weight:** | | | | | |
| Sulfur | 0.01 to 0.5 | 0.05 to 1.0 | 0.2 to 2.0 | 0.5 to 3.0 | 0.7 to 3.5 |
| Hydrogen | 13.3 to 14.1 | 11.8 to 13.9 | (10.6 to 13.0)* | (10.5 to 12.0)* | (9.5 to 12.0)* |
| Carbon | 85.9 to 86.7 | 86.1 to 88.2 | (86.5 to 89.2)* | (86.5 to 89.2)* | (86.5 to 90.2)* |
| Nitrogen | nil to 0.1 | nil to 0.1 | ---- | ---- | ---- |
| Oxygen | — | — | | | |
| Ash | ---- | ---- | 0 to 0.1 | 0 to 0.1 | 0.01 to 0.5 |
| **Gravity:** | | | | | |
| Deg API | 40 to 44 | 28 to 40 | 15 to 30 | 14 to 22 | 7 to 22 |
| Specific | 0.825 to 0.806 | 0.887 to 0.825 | 0.966 to 0.876 | 0.972 to 0.922 | 1.022 to 0.922 |
| lb/gal | 6.87 to 6.71 | 7.39 to 6.87 | 8.04 to 7.30 | 8.10 to 7.68 | 8.51 to 7.68 |
| Pour point, F | 0 to -50 | 0 to -40 | -10 to +50 | -10 to +80 | +15 to +85 |
| **Viscosity:** | | | | | |
| Centistokes at 100F | 1.4 to 2.2 | 1.9 to 3.0 | 10.5 to 65 | 65 to 200 | 260 to 750 |
| SUS at 100F | — | 32 to 38 | 60 to 300 | — | — |
| SSF at 122F | — | — | — | 20 to 40 | 45 to 300 |
| Water and sediment, % by vol | — | 0 to 0.1 | tr to 1.0 | 0.05 to 1.0 | 0.05 to 2.0 |
| Heating value, Btu/lb gross (calculated) | 19,670 to 19,860 | 19,170 to 19,750 | 18,280 to 19,400 | 18,100 to 19,020 | 17,410 to 18,990 |

*Estimated

**Table 3-7:** Properties of Fuel Oils (Reproduced with permission from *Steam*, 40th ed., published by Babcock & Wilcox, a McDermott Company)

| **Property** | **Definition** |
|---|---|
| API Gravity | Developed by the American Petroluem Institute to determine the relative density of oils. Lower values indicate higher density. |
| Flash Point | Temperature to which a liquid must be heated to produce vapors that flash but do not burn continuously when ignited. This is different than the ignition temperature, which is the temperature which ingites vapors. |
| Pour Point | The lowest temperature at which a liquid fuel flows under atmospheric conditions. |
| Viscosity | Measure of a liquid's internal resistance to flow. |

**Table 3-8:** Definitions of fuel oil properties (Source: *Steam*, 40th ed., published by Babcock & Wilcox, a McDermott Company, and Sinder, J.G., ed., *Combustion: Fossil Power*, Alstom)

# NATURAL GAS

Natural gas is found in independent pockets or in oil or coalfields. The prime component of natural gas is methane ($CH_4$), although as Table 3-9 illustrates, other chemicals are present, most notably ethane ($C_2H_6$), which is the lightest hydrocarbon next to methane. Natural gas has found historic favor as a primary fuel for power production because it is easy to handle, has a pipeline infrastructure in place, burns cleanly, and produces fewer pollutants than other fuels when burned. Of the three major steam generation fossil fuels, natural gas has the highest heating value per unit weight. This aspect and recent technology advancements have led to the proliferation of gas-powered combined-cycle units, some of which

| Sample No. | 1 | 2 | 3 | 4 | 5 |
|---|---|---|---|---|---|
| Source: | Pa. | S.C . | Ohio | La. | Ok. |
| Analyses: | | | | | |
| Constituents, % by vol | | | | | |
| $H_2$, Hydrogen | -- | -- | 1.82 | -- | -- |
| $CH_4$, Methane | 83.40 | 84.00 | 93.33 | 90.00 | 84.10 |
| $C_2H_4$,Ethylene | -- | 0.25 | -- | -- | -- |
| $C_2H_6$, Ethane | 15.80 | 14.80 | -- | 5.00 | 6.70 |
| CO, Carbon monoxide | -- | -- | 0.45 | -- | -- |
| $CO_2$, Carbon dioxide | -- | 0.70 | 0.22 | -- | 0.80 |
| $N_2$, Nitrogen | 0.80 | 0.50 | 3.40 | 5.00 | 8.40 |
| $O_2$, Oxygen | -- | -- | 0.35 | -- | -- |
| $H_2S$, Hydrogen sulfide | -- | -- | 0.18 | -- | -- |
| Ultimate, % by wt | | | | | |
| S, Sulfur | -- | -- | 0.34 | -- | -- |
| $H_2$, Hydrogen | 23.53 | 23.30 | 23.20 | 22.68 | 20.85 |
| C, Carbon | 75.25 | 74.72 | 69.12 | 69.26 | 64.84 |
| $N_2$, Nitrogen | 1.22 | 0.76 | 5.76 | 8.06 | 12.90 |
| $O_2$, Oxygen | -- | 1.22 | 1.58 | -- | 1.41 |
| Specific gravity (rel to air) | 0.636 | 0.636 | 0.567 | 0.600 | 0.630 |
| HHV | | | | | |
| Btu/ft³ at 60F and 30 in. Hg | 1,129 | 1,116 | 964 | 1,022 | 974 |
| (kJ/m³ at 16C and 102 kPa) | (42,065) | (41,581) | (35,918) | (38,079) | (36,290) |
| Btu/lb(kJ/kg) of fuel | 23,170 (53,893) | 22,904 (53,275) | 22,077 (51,351) | 21,824 (50,763) | 20,160 (46,892) |

**Table 3-9:** Analyses of several natural gas supplies in the U.S. (Reproduced with permission from *Steam*, 40th ed., published by Babcock & Wilcox, a McDermott Company)

operate at near 60% efficiency. This is almost twice the efficiency of the best coal-fired plant. Because natural gas as delivered is a very clean fuel, combustion products (ash, slag) are non-existent.

Coal, natural gas, and oil comprise the major fossil fuels for steam generation. Other minor fuels include wood, coke, refuse, coke oven gas, and blast furnace gas.

## Ash properties and effects in the boiler

A quote from *Combustion: Fossil Power* (J. G. Snider, ed., Alstom, 1991) sums up the difficulties with solid combustion byproducts: "Without ash, all furnaces could easily be designed on the basis of heat transfer only."

Inorganic minerals contained within fuel do not burn with the combustion process, and either transport out of the boiler or deposit on boiler internals. Deposition may consist of molten and partially-molten compound buildups on furnace tubes (slagging) or ash accumulation on convective pass superheater and reheater tubes (fouling). The volume and complexity of ash are obviously greatest in a coal-fired unit, and in large measure coal plants are designed around ash characteristics and removal requirements.

Table 3-5 outlines the fuel analysis from coals around the U.S., but it also illustrates the ash chemistry of these coals. The data reveal several interesting details:

- The ash content of the two western subbituminous coals is lower than that of all other coals, although lower ash content does not necessarily mean less fouling

- All coals contain significant amounts of silica and aluminum. These come from the complex aluminosilicates that comprise much of the earth's crust

- The variable content of iron, the alkalis sodium and potassium, and the alkaline earth minerals, calcium and magnesium, greatly influence ash melting temperatures and other properties, which in turn influence slagging and fouling

- The variable sulfur concentrations—besides the air pollution aspects, sulfur compounds play a direct role in boiler tube corrosion

The reader should note that ash constituents are all reported as oxides. This is the standard method for reporting ash analyses, but as seen in Table 3-3, the original minerals are usually more complex. As a lead-in to an examination of ash chemistry and its effect upon boilers, a first concept to consider is ash melting behavior (commonly known as fusibility) and its importance in predicting the behavior of ash in boilers. The ASTM has developed a test for determining the melting characteristics of ash. The test involves placing an ash sample into a small pyramid, subjecting it to controlled heating, and measuring deformation characteristics. The four parameters are:

- initial deformation temperature (IT)

- softening temperature (SD)

- hemispherical temperature (HT)

- fluid temperature (FT)

---

| **Definition of Ash Fusion Criteria** | |
| --- | --- |
| Initial Deformation Temperature (IT) | The temperature at which the tip of the ash pyramid begins to show any evidence of deformation. |
| Softening Temperature (ST) | The temperature at which the sample has fused into a spherical shape where the height is equal to the width of the base, H = W. The ST temperature in a reducing atmosphere is often referred to as the "fusion temperature." |
| Hemispherical Temperature (HT) | The temperature at which the sample has fused into a hemispherical shape where the height is equal to the 1/2 the width of the base. |
| Fluid Temperature (FT) | The temperature at which the sample has fused into a nearly flat layer with a maximum height of 1/16 inch. |

**Table 3-10:** Definition of ash fusion criteria (Source: Sinder, J.G., ed., *Combustion: Fossil Power*, Alstom)

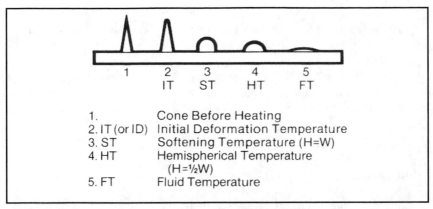

1.                Cone Before Heating
2. IT (or ID)   Initial Deformation Temperature
3. ST         Softening Temperature (H=W)
4. HT         Hemispherical Temperature
                    ($H=\frac{1}{2}W$)
5. FT         Fluid Temperature

**Fig. 3-1:** Illustration of fusion temperatures (Reproduced with permission from *Combustion: Fossil Power*, published by Alstom)

| Element | Oxide | Melting Temp (°F) | Chemical Property |
|---------|-------|-------------------|-------------------|
| Si | $SiO_2$ | 3120 | Acidic |
| Al | $Al_2O_3$ | 3710 | Acidic |
| Ti | $TiO_2$ | 3340 | Acidic |
| Fe | $Fe_2O_3$ | 2850 | Basic |
| Ca | CaO | 4570 | Basic |
| Mg | MgO | 5070 | Basic |
| Na | $Na_2O$ | Sublimes at 2330 | Basic |
| K | $K_2O$ | Decomposes at 660 | Basic |

**Table 3-11:** Melting temperatures of simple materials (Reproduced with permission from *Combustion: Fossil Power*, published by Alstom)

| Compound | Melting Temp (°F) |
|----------|-------------------|
| $Na_2SiO_3$ | 1610 |
| $K_2SiO_3$ | 1790 |
| $Al_2O_3 \cdot Na_2O \cdot 6SiO_2$ | 2010 |
| $Al_2O_3 \cdot K_2O \cdot 6SiO_2$ | 2100 |
| $FeSiO_3$ | 2090 |
| $CaO \cdot Fe_2O_3$ | 2280 |
| $CaO \cdot MgO \cdot 2SiO_2$ | 2535 |
| $CaSiO_3$ | 2804 |

**Table 3-12:** Melting temperatures of complex minerals found in coal ash (Reproduced with permission from *Combustion: Fossil Power*, published by Alstom)

The definition of these temperatures is illustrated in Table 3-10 and Figure 3-1.

Table 3-4 shows that fusibility temperatures may vary markedly with a change from oxidizing to reducing atmospheres. In large part this is due to the conversion of iron compounds in the coal to various species of different oxidation states. More details on this aspect will appear later.

# Slagging

Slag formation is directly influenced by the fusibility properties of ash. The combustion zone of the boiler is the highest heat area, and ash is often molten in this region. When completely molten, ash flows readily. However, the viscosity can

| Mineral Relationships Important to Ash Fusion Temperatures | |
|---|---|
| Base/acid ratio = | $\dfrac{Fe_2O_3+CaO+MgO+Na_2O+K_2O}{SiO_2+Al_2O+TiO}$ |
| Silica/alumina ratio= | $\dfrac{SiO_2}{Al_2O_3}$ |
| Iron/calcium ratio= | $\dfrac{Fe_2O_3}{CaO}$ |
| Iron/dolomite ratio= | $\dfrac{Fe_2O_3}{CaO+MgO}$ |
| Dolomite percentage (DP)= | $\dfrac{CaO+MgO}{CaO+MgO+Fe_2O_3+Na_2O+K_2O}$ |
| Equivalent ferric oxide (equiv. $Fe_2O_3$)= | $Fe_2O_3+1.11\ FeO+1.43\ Fe$ |
| Silica Percentage= | $\dfrac{SiO_2}{SiO_2+Equiv.Fe_2O_3+CaO+MgO}$ |
| Total alkalis= | $Na_2O+K_2O$ |

**Table 3-13:** Mineral relationships important to ash fusion temeratures (Source: *Combustion: Fossil Power*, published by Alstom)

increase dramatically in some ashes with just a small temperature drop. Ash particles between IT and HT conditions can form very sticky particles that bond tightly to tube walls. Heavy slagging reduces heat transfer in the waterwalls, which increases the furnace exit gas temperature (FEGT). Higher FEGT may cause unwanted steam temperature increases and may extend slag formation into further reaches of the boiler.

Each of the different minerals in ash has an effect upon fusibility. The melting points of some of the individual minerals are shown in Table 3-11. Interactions between minerals very much complicate the issue and combinations of compounds may dramatically lower fusion temperatures, as Table 3-12 indicates. Many other factors influence fusibility and the discussion can become rather complicated, so some simplification is warranted. Table 3-13 outlines in equation form the important influences minerals have in coal ash. The following discussion explains these terms:

**Base/acid ratio.** Some of the minerals that form during coal combustion are considered to be basic in nature, and some are considered acidic. The basic minerals are the iron, calcium, magnesium, sodium, and potassium oxides. The acidic

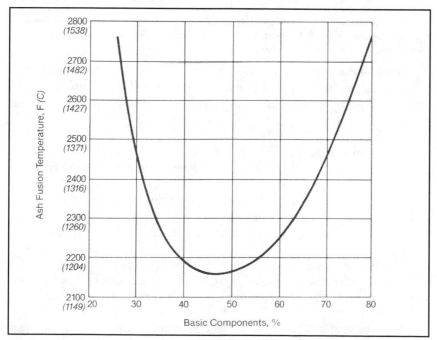

**Fig. 3-2:** Ash fusion temperatures as a function of base/acid ratio (Reproduced with permission from *Steam*, 40th ed., published by Babcock & Wilcox, a McDermott Company)

minerals are silica, alumina, and titanium dioxide. The ratio of basic to acidic minerals in ash greatly influences melting temperatures. Figure 3-2 illustrates the concept graphically. Base/acid ratios at or near 45% typically produce the lowest melting temperatures.

**Silica/alumina ratio.** The ratio of silica to alumina also influences melting temperatures. When mixed with basic partners, silica tends to produce lower ash melting temperatures than if the acidic component was alumina.

**Iron/calcium ratio.** Another important parameter is the iron/calcium ratio, where the ferric oxide ($Fe_2O_3$) portion of iron is taken into account. The fluxing action between iron and calcium is rather complex. Snider, in *Combustion: Fossil Power*, describes the interactions in greater detail. Figure 3-3 shows the general trend, which is that lower $Fe_2O_3/CaO$ ratios tend to lower the ash softening temperature.

| Coal | $Fe_2O_3$ | CaO | $Fe_2O_3/CaO$ Ratio | Softening Temperature (°F) |
|------|-----------|-----|---------------------|----------------------------|
| (1) | 31.8 | 0.3 | 106.0 | 2,360 |
| (2) | 24.8 | 2.0 | 12.4 | 2,270 |
| (3) | 21.3 | 4.8 | 4.4 | 2,130 |

**Fig. 3-3:** Influence of iron/calcium ratios on fusion temperatures (Reproduced with permission from *Combustion: Fossil Power*, published by Alstom)

**Iron/dolomite ratio.** Dolomite is a common form of limestone, in which calcium is supplemented to a great extent by magnesium. The iron/dolomite ratio impacts ash-softening temperatures in much the same way as the iron/calcium ratio.

**Dolomite percentage.** With two coals of a similar base content, the one with the higher percentage of dolomite will tend to have higher fusion temperatures.

**Equivalent ferric oxide.** This definition explains the effects of iron in oxidizing or reducing conditions. In a boiler fired with excess air at the main burners, most of the iron in the coal reacts to form ferric oxide.

However, in a reducing atmosphere—such as in a boiler with overfire air for $NO_x$ control—the iron in the reducing zone will contain some FeO and perhaps some metallic iron (Fe). These compounds lower ash-melting temperatures.

**Silica percentage.** As silica percentage increases compared to the other fluxing agents—iron, calcium oxide, and magnesium oxide—the ash viscosity increases.

**Total alkalis.** Sodium and potassium lower ash-melting temperatures. Alkalis are important with respect to fouling, as is discussed in the next section.

A wide variety of factors influence slagging properties. It is impossible to completely predict the slagging properties of a coal, although a set of calculations for predicting slagging potential can be found in Babcock & Wilcox, *Steam*, 40th ed., (1992). Some general observations are evident.

Slagging is problematic if ash particles contact waterwall tubes while the particles are in a partially molten state. This is the range between the IT and HT temperatures defined earlier (sometimes defined as "the plastic region"). Ash with a wide temperature variation between the IT and HT will be plastic for an extended period of time. For bituminous ash, the slagging potential is directly related to the base/acid ratio and sulfur content. With lignitic ash, the slagging potential is essentially a function of the temperature difference between IT and HT. The bituminous and lignitic slagging calculations are outlined extensively in Babcock & Wilcox. As Table 3-14 illustrates, the change from an oxidizing to reducing atmosphere substantially affects fusibility temperatures. In most cases, this is due to iron and the varied species that form in reduced atmospheres. In contrast, the change in atmospheres does not substantially alter the fusion temperatures for the subbituminous coal in the table. This is a lignitic ash coal with a small amount of iron.

Slag control is readily apparent in boiler design. Cyclone units, popular in the 1960s, were designed to produce molten slag within the cyclone barrels and lower

| Rank: | Low Volatile Bituminous | High Volatile Bituminous | | | | Sub-bituminous | Lignite |
|---|---|---|---|---|---|---|---|
| Seam | Pocahontas No. 3 | No. 9 | No.6 | Pittsburgh | | Antelope | |
| Location | West Virginia | Ohio | Illinois | West Virginia | Utah | Wyoming | Texas |
| Ash, dry basis,% | 12.3 | 14.1 | 17.4 | 10.9 | 17.1 | 6.6 | 12.8 |
| Sulfur, dry basis, % | 0.7 | 3.3 | 4.2 | 3.5 | 0.8 | 0.4 | 1.1 |
| Analysis of ash, % by wt | | | | | | | |
| $SiO_2$ | 60.0 | 47.3 | 47.5 | 37.6 | 61.1 | 28.6 | 41.8 |
| $Al_2O_3$ | 30.0 | 23.0 | 17.9 | 20.1 | 21.6 | 11.7 | 13.6 |
| $TiO_2$ | 1.6 | 1.0 | 0.8 | 0.8 | 1.1 | 0.9 | 1.5 |
| $Fe_2O_3$ | 4.0 | 22.8 | 20.1 | 29.3 | 4.6 | 6.9 | 6.6 |
| CaO | 0.6 | 1.3 | 5.8 | 4.3 | 4.6 | 27.4 | 17.6 |
| MgO | 0.6 | 0.9 | 1.0 | 1.3 | 1.0 | 4.5 | 2.5 |
| $Na_2O$ | 0.5 | 0.3 | 0.4 | 0.8 | 1.0 | 2.7 | 0.6 |
| $K_2O$ | 1.5 | 2.0 | 1.8 | 1.6 | 1.2 | 0.5 | 0.1 |
| $SO_3$ | 1.1 | 1.2 | 4.6 | 4.0 | 2.9 | 14.2 | 14.6 |
| $P_2O_5$ | 0.1 | 0.2 | 0.1 | 0.2 | 0.4 | 2.3 | 0.1 |
| Ash fusibility | | | | | | | |
| Initial deformation temp, F | | | | | | | |
| Reducing | 2900 + | 2030 | 2000 | 2030 | 2180 | 2280 | 1975 |
| Oxidizing | 2900 + | 2420 | 2300 | 2265 | 2240 | 2275 | 2070 |
| Softening temp, F | | | | | | | |
| Reducing | | 2450 | 2160 | 2175 | 2215 | 2290 | 2130 |
| Oxidizing | | 2605 | 2430 | 2385 | 2300 | 2285 | 2190 |
| Hemispherical temp, F | | | | | | | |
| Reducing | | 2480 | 2180 | 2225 | 2245 | 2295 | 2150 |
| Oxidizing | | 2620 | 2450 | 2450 | 2325 | 2290 | 2210 |
| Fluid temp, F | | | | | | | |
| Reducing | | 2620 | 2320 | 2370 | 2330 | 2315 | 2240 |
| Oxidizing | | 2670 | 2610 | 2540 | 2410 | 2300 | 2290 |

**Table 3-14:** Ash content and fusion temperatures of some U.S. coals (Reproduced with permission from *Steam*, 40th ed., published by Babcock & Wilcox, a McDermott Company)

portions of the boiler, and drain it as a liquid to the water-filled slag tank. This is the wet-bottom concept (that stands for the molten slag, not the water-filled slag tank) mentioned in chapter 1. The ratio of bottom ash to flyash in a cyclone unit is on the order of 80:20.

Most pulverized coal units operate differently. Tiny coal particles burn quickly, and the fine ash residue is carried upward with furnace flow. In a properly designed system, ash particles that contact the furnace walls have already solidified and do not stick. Ash ratios are almost always reversed in these types of units—80% escapes as flyash and 20% as bottom ash. Because the bottom ash does not discharge in a molten state, these are known as dry-bottom units.

Figure 3-4 shows relative boiler sizes with coals of different slagging properties for a 660-MW application. This figure is rather revealing, especially when one considers the popular phenomenon of fuel switching. Whereas the boiler may have been designed to handle a low-slagging coal, it might now be fired with a higher-slagging fuel. Several remedies are possible to mitigate the change in slagging potential. These include installation of additional sootblowers, firing at reduced rates, and chemical additives to improve fuel properties. Appendix 3-2 outlines the case history of a utility that switched from bituminous coal to PRB coal for firing of a cyclone and pulverized-coal boiler.

# Fouling

Fouling is most prominent in the convection pass of the boiler, primarily in the superheater and reheater areas. Fouling is caused by the deposition of flyash

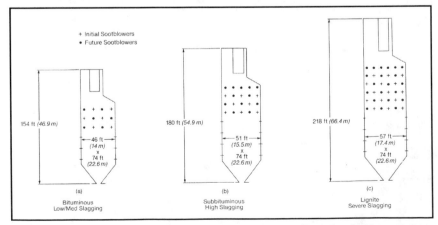

**Fig. 3-4:** Relative boiler sizing as a function of slagging properties (Reproduced with permission from *Steam*, 40th ed., published by Babcock & Wilcox, a McDermott Company)

particles on tube and duct surfaces. In the absence of furnace upsets, the ash particles that enter the convection pass are in solid form, and by themselves should not have a strong propensity to stick to equipment. However, combustion generates volatile alkali compounds of sodium and potassium, which condense on tube surfaces and ash particles, giving them much stronger adhesion tendencies.

The concentration of sodium and potassium in the flue gas is directly related to the manner in which the two elements are bound within the original fuel. Sodium and potassium combined with silicates tend to remain with these compounds throughout the process. However, a percentage of the alkalis either exist as simple salts (primarily chlorides) or are organically bound in the coal. These "active" alkalis vaporize during the combustion process and form the oxides $Na_2O$ and $K_2O$. The relatively lower temperatures in the convective pass of the boiler allow the alkalis to condense.

Two interrelated aspects of fouling are most critical in boiler design:

- The fouling potential of the coal or, more properly, the alkali residue

- The nature of the deposits that accumulate and how easily they may be removed by sootblowing and other mechanical methods

| Fouling Potential | Cholorine(%) | Equivalent Sodium (%) |
|---|---|---|
| High | >0.5 | >0.33 |
| Medium | 0.3 | 0.2 |
| Low | <0.1 | <0.7 |

**Table 3-15**: Fouling tendencies as related to coal chlorine content (Reproduced with permission from *Combustion: Fossil Power*, published by Alstom)

With respect to fouling potential, the bulk of active alkalis exist as chloride salts, so a measure of coal chlorine content is a standard guideline to determine the relative fouling potential (Table 3-15). The company that produced these data developed its own procedure for determining active alkali, which involves washing the coal sample with a weak acid to extract soluble alkalis. While the details of this procedure do not need to be outlined here, the results are revealing (Table 3-16). The data illustrate the fouling tendencies of several U.S. coals. As is clearly evident, the fouling potential is directly related to the sodium content of the fuel. Of particular note—especially to utilities that have switched fuel to reduce $SO_2$ emissions—is the difference in fouling potential between the two bituminous coals and the subbituminous coal from Montana. The sodium content of the bituminous coals is half that of the subbituminous coal.

| Rank Region | Lignite ND | Sub B MT | Lignite TX (Yegua) | Lignite TX (Wilcox) | hvBb UT | hvAb PA | Lignite TX (Wilcox) |
|---|---|---|---|---|---|---|---|
| HHV, Btu/lb, Dry Basis | 10640 | 12130 | 7750 | 9710 | 12870 | 13200 | 8420 |
| **Ash Composition (%)** | | | | | | | |
| $SiO_2$ | 20.0 | 33.9 | 62.1 | 52.3 | 52.5 | 51.1 | 57.9 |
| $Al_2O_3$ | 9.1 | 11.4 | 15.1 | 17.4 | 18.9 | 30.7 | 21.8 |
| $Fe_2O_3$ | 10.3 | 10.8 | 3.5 | 5.3 | 1.1 | 10.0 | 3.9 |
| CaO | 22.4 | 21.0 | 6.2 | 9.4 | 13.2 | 1.6 | 7.1 |
| MgO | 6.4 | 2.7 | 0.7 | 3.2 | 1.3 | 0.9 | 2.1 |
| $Na_2O$ | 5.0 | 5.8 | 3.6 | 0.9 | 3.8 | 0.4 | 0.7 |
| $K_2O$ | 0.5 | 1.6 | 1.9 | 1.2 | 0.9 | 1.7 | 0.8 |
| TiO | 0.4 | 0.7 | 0.9 | 1.2 | 1.2 | 2.0 | 1.1 |
| $SO_3$ | 21.9 | 12.0 | 6.1 | 9.6 | 6.2 | 1.4 | 4.4 |
| **Fouling Potential** | Severe | High | High | Moderate | Moderate | Low | Low |
| **Lb Ash/$10^6$ Btu, Dry Basis** | 9.0 | 4.6 | 43.3 | 20.1 | 7.9 | 10.2 | 34.4 |
| **Acetic-Acid-Soluble** | | | | | | | |
| Sodium (Na, ppm) | 3980 | 2680 | 9650 | 1030 | 1120 | 250 | 340 |
| Potassium (K, ppm) | . . . | . . . | 1230 | 85 | 85 | . . . | 110 |
| **Alkali in Ash, % Wt.** | | | | | | | |
| $Na_2O$ | 5.0 | 5.8 | 3.6 | 0.9 | 3.8 | 0.4 | 0.7 |
| $K_2O$ | 0.5 | 1.6 | 1.9 | 1.2 | 0.9 | 1.7 | 0.8 |
| **Equiv. Sol. Alkali in Ash, % Wt. of Ash** | | | | | | | |
| $Na_2O$ | 5.58 | 6.45 | 3.88 | 0.71 | 1.49 | 0.15 | 0.16 |
| $K_2O$ | . . . | . . . | 0.44 | 0.04 | 0.08 | . . . | 0.05 |
| **% Sol. Alkali of Total (Equiv. Sol. $Na_2O$)/ ($Na_2O$ in Ash)** | 112 | 111 | 108 | 79 | 39 | 38 | 23 |
| **(Lbs Sol. Na)/ ($10^6$ Btu Fired)** | 0.374 | 0.221 | 1.245 | 0.106 | 0.087 | 0.018 | .040 |
| **(Lbs Sol. Na)/(lb Ash/ $10^6$ Btu Fired)** | 0.044 | 0.048 | 0.223 | 0.005 | 0.014 | 0.002 | 0.001 |

**Table 3-16:** Alkali content of some U.S. coals (Reproduced with permission from *Combustion: Fossil Power*, published by Alstom)

The second relationship between sodium and fouling is that greater active sodium content increases the strength of ash deposits. Figure 3-5 shows that this is true for both bituminous ash and lignitic ash coals. In both cases, but particularly with lignitic ash, ash strength increases enormously with increasing sodium oxide content.

Fouling of convective pass equipment and surfaces causes a number of problems. Obviously, buildups on superheater tubes reduce heat transfer efficiency. But buildups also cause channeling (or "laning") of the flue gas, which increases linear velocity through open areas. The increased velocity may then increase ash erosion of other tubes. It is not uncommon for tube failures to be caused by ash erosion. Excessive ash buildups between superheater pendants may cause bridging of material between pendant sections.

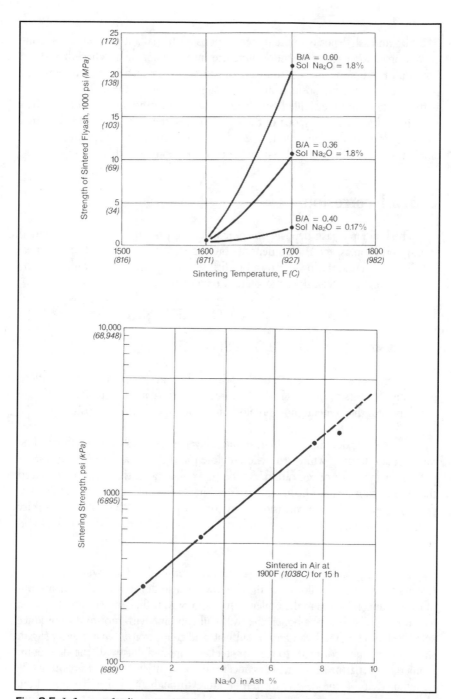

**Fig. 3-5:** Influence of sodium concentration on sintered ash strength (Reproduced with permission from *Steam*, 40th ed., published by Babcock & Wilcox, a McDermott Company)

Slag and ash deposits in the upper sections of the radiant portion of the boiler may break loose and damage or puncture furnace floor tubes when they strike bottom. In an unusual but not unheard-of scenario, personnel at one utility would periodically blast ash deposits from superheater tubes (while the unit was on-line) with a shotgun and shells filled with bird seed or some other material lighter than lead shot. However, during one of these procedures, the blast caused a failure in a superheater tube. Was this due to the tube's weakened condition, or from corrosion? It was never determined. Safety concerns put an end to this procedure.

## Coal ash corrosion

The buildup of ash deposits in the convective pass may also be detrimental from a corrosion aspect. As ash deposits build up, volatile alkalis and sulfur trioxide ($SO_3$) produced during combustion diffuse through the ash to initiate corrosion reactions. Two typical reactions are shown below.

$$3K_2SO_4 + Fe_2O_3 + 3SO_3 \rightarrow 2K_3Fe(SO_4)_3$$

$$K_2SO_4 + Al_2O_3 + 3SO_3 \rightarrow 2KAl(SO_4)_2$$

Whereas sodium appears to be the prime culprit in causing deposit formation, potassium appears to be the chief alkali initiating the corrosion reactions. Figure 3-6 illustrates the "general morphology" of a coal ash corrosion deposit.

Three layers develop. The first, an outer layer, is essentially just the flyash. The intermediate layer is white- to yellow-colored and shows a marked increase in potassium and $SO_3$ concentration. As the figure illustrates, this layer has replaced the original tube metal. The inner layer—a thin black band located at the tube surface—is the site of active corrosion. Iron content is high because of its proximity to the base metal, and the compounds within this layer include iron sulfides and sulfates.

Low-temperature corrosion of air heaters and outlet ducts will occur if the flue gas temperature is allowed to drop below the acid dew point. A small amount of the sulfur combusted in the boiler converts to $SO_3$. If the temperature drops too low at the backend of the boiler, the $SO_3$ will combine with moisture to produce sulfuric acid ($H_2SO_4$). Although the sulfuric acid concentration may be very slight, the liquid is quite corrosive to carbon steel. Figure 3-7 illustrates the dew point temperature as a function of $SO_3$ concentration. Exit gas temperatures must be maintained above the dew point temperature through the boiler backpass (and electrostatic precipitator, if the unit has one) to prevent corrosion.

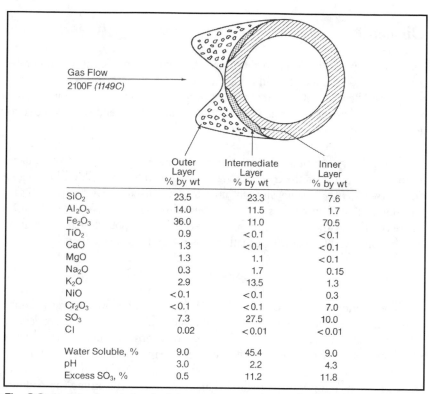

| | Outer Layer % by wt | Intermediate Layer % by wt | Inner Layer % by wt |
|---|---|---|---|
| $SiO_2$ | 23.5 | 23.3 | 7.6 |
| $Al_2O_3$ | 14.0 | 11.5 | 1.7 |
| $Fe_2O_3$ | 36.0 | 11.0 | 70.5 |
| $TiO_2$ | 0.9 | <0.1 | <0.1 |
| CaO | 1.3 | <0.1 | <0.1 |
| MgO | 1.3 | 1.1 | <0.1 |
| $Na_2O$ | 0.3 | 1.7 | 0.15 |
| $K_2O$ | 2.9 | 13.5 | 1.3 |
| NiO | <0.1 | <0.1 | 0.3 |
| $Cr_2O_3$ | <0.1 | <0.1 | 7.0 |
| $SO_3$ | 7.3 | 27.5 | 10.0 |
| Cl | 0.02 | <0.01 | <0.01 |
| | | | |
| Water Soluble, % | 9.0 | 45.4 | 9.0 |
| pH | 3.0 | 2.2 | 4.3 |
| Excess $SO_3$, % | 0.5 | 11.2 | 11.8 |

**Fig. 3-6:** Analysis of a typical coal ash deposit from a superheater tube (Reproduced with permission from *Steam*, 40th ed., published by Babcock & Wilcox, a McDermott Company)

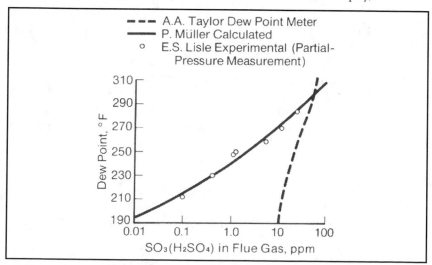

**Fig. 3-7:** Influence of $SO_3$ concentration on the acid dew point (Reproduced with permission from *Steam*, 40th ed., published by Babcock & Wilcox a McDermott Company)

# Oil ash

Oil contains much less mineral content than coal, so oil ash deposition problems are not as complicated. Oil ash typically does not cause corrosion of water-wall tubes. The problem areas are the superheater and reheater due to low-melting ash deposits. Also, as with coal combustion products, back-end corrosion due to acid dew point corrosion is a possibility.

The primary culprit in ash corrosion is vanadium, which comes from minerals in the soil and from the creatures that decomposed to form oil. Oils throughout the world may contain virtually no vanadium to almost 400 ppm (mg/l). Vanadium released in combustion forms several oxides, $V_2O_3$, $V_2O_4$, and $V_2O_5$. These combine with alkali salts to form low-melting compounds that accumulate on tube surfaces. The following equation illustrates a typical reaction:

$$Na_2SO_4 + V_2O_5 \rightarrow 2NaVO_3 + SO_3$$

The melting point of $NaVO_3$ is 1,165°F (629°C). The low-melting sodium-vanadium compounds are directly corrosive to steel. The corrosion rate accelerates with increasing metal temperature. Methods of control include selecting low-vanadium oil, boiler design to reduce metal temperature, good sootblower arrangement to keep tubes clean, and use of chemical additives to control corrosion. Common additives include alumina and magnesium containing compounds. These increase the melting point of ash deposits.

# Conclusion

This chapter illustrates that for coal-fired power generation and to a lesser extent oil-fired generation, ash properties play a critical part in design and operation of the boiler. Ash characteristics are also critical at coal-fired plants that switch fuels to lower flue gas emissions or fuel costs. These retrofit applications always require boiler modifications, such as installation of new sootblowers, to handle the change in ash properties.

# Appendix 3-1

## PRB Coal Switch Not A Complete Panacea

### Brad Buecker
### John Meinders

Low-sulfur coal from the Powder River Basin in Wyoming is being used by utilities to reduce sulfur dioxide emissions. However, the changeover to Powder River Basin (PRB) coal, in units originally designed for other coals, introduces a number of complications. Some of the problems relate to the general characteristics of the fuel and some to its combustion behavior in the furnace.

## Fuel switch

In 1998, the Kansas City, Kansas Board of Public Utilities' Quindaro Power Station switched from Midwestern coals to PRB. The Quindaro power station is located along the Missouri River in Kansas City, Kansas. At Quindaro, the primary generating sources are an 82 MW Babcock & Wilcox cyclone unit (Unit No. 1), commissioned in 1966, and a 145 MW, Riley Stoker, pulverized coal (PC) unit (Unit No. 2), commissioned in 1971. The PC unit was retrofitted with Combustion Engineering burners and dampers in 1992, while the cyclone boiler was converted from forced-draft to balanced-draft operation in the same year.

When originally put into operation, both units burned a high-sulfur coal from southern Kansas. However, in 1979 the fuel was changed to an Illinois coal, which remained the fuel of choice until December of 1998, when both units started burning PRB coal exclusively. The fuel switch required several modifications to the units.

# Coal quality and handling

The original coal handling and preparation system included: a rail car unloading (bottom dumping) facility, a belt conveyor system to a stack-out pile, a Pennsylvania Crusher granulator and hammer mill crusher for the cyclone unit, three coal storage silos, and three ball mills for the PC unit.

PRB coals differ from their eastern counterparts in several respects. PRB coals have a lower heat content, lower ash content (but with a different mineral chemical composition particularly high in calcium), higher moisture, and a greater tendency for spontaneous combustion.

Spontaneous combustion requires special attention, especially in the coal handling area. In preparation for using PRB coal, the Quindaro staff installed a dry-pipe sprinkler for the entire fuel handling system and coal silos. Furthermore, all of the fuel handling electrical equipment was upgraded to meet NFPA and IEEE codes. Quindaro also installed BetzDearborn dust suppression systems at the stack-out pile and coal fines/dust collectors.

A coal fines removal system, manufactured by Air Cure, services the complete fuel handling system. The system, which operates under a slight vacuum, has air emission vents located at the coal unloading facility, the coal silos and bunkers, and all transfer points. A series of reverse pulse baghouses collects the fines.

After collection, the fines are pneumatically conveyed to the operating coal bunker(s). When the PC unit is not in service, the fines are blended with the primary fuel. It is important that fines not be conveyed to idle bunkers, as excess fines can cause combustion problems when they are introduced to the cyclone boiler.

The Quindaro power station's baghouses have been equipped with sprinkler systems. However, even after the sprinklers have been used, burning fines within the baghouse can still occur. PRB coal fires in baghouses are a big problem and if not corrected can cause serious damage to the equipment and injury to the operators.[1]

The operators also periodically wash the coal handling facility with water. All of the wash water effluent flows to a containment pond, which is periodically decanted by the operators to the plant waste treatment system. An operator then mechanically transports the fines back to the coal pile.

On the recommendation of Pennsylvania Crusher, the plant replaced the coal hammer mill crusher with a grate-type cage crusher, which is reported to be better at handling higher moisture PRB coal. Water spray and steam inerting systems have been installed in the ball mills to deal with emergency situations.

For saftey reasons, the operators use the steam inerting system during startup and shutdown of the mills, which are the most critical periods. An initial drawback of using this method was that steam entered the burners and caused problems with lighting of the fuel. To overcome this, BPU installed separate igniters.

## Boiler issues

The most difficult issues to solve have been with the cyclone unit. One of the major factors that made cyclone boilers popular a few decades ago was their combustion efficiency. Combustion of bituminous coals in a cyclone produces a coat of molten slag on the cyclone walls. The slight buildup of slag increases the residence time in the furnace and thus provides for nearly complete combustion.

Figure A3-1 shows a comparison of the temperature versus viscosity relationship between an Illinois coal and a PRB coal. As is evident, the PRB coal goes from a high viscosity to a very low viscosity over a short temperature range. However, it is still below the normal operating temperature of the cyclone boiler. The result is a thin layer of runny slag that reduces the residence time of the fuel in the furnace. This allows fuel to blow through the cyclone, which increases flyash content and the amount of unburned carbon in the flyash. In addition to reducing fuel efficiency, unburned carbon increases the potential for precipitator fires.

To overcome this problem, Quindaro management initially installed a baffle in the secondary air stream on each of the unit's two cyclones. Although this modified the airflow and increased the fuel residence time, it did not completely solve

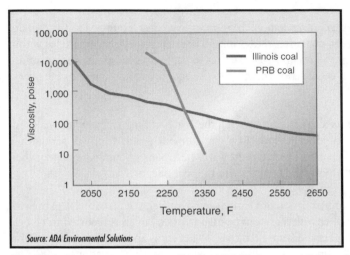

**Fig. A3-1:** Slag viscosity comparison (Source: ADA Environmental Solutions)

the problem. In recent months, the staff has been working with ADA Environmental Solutions on a chemical solution.

ADA has performed a series of tests that involve the feed of a proprietary mixture of iron oxides and stabilizing chemicals into the coal before combustion. The chemical mixture enhances the viscosity characteristics of the slag. Preliminary results at Quindaro[2,3] have been very promising.

Prior to the tests, unit load could not be reduced below 75% without slag freeze-ups in the boiler bottom. Now, with a 1% mixture of the chemical, the boiler load can be reduced to 47%. Video inspections indicate improved slag formation and more complete combustion within the cyclone. This is reflected by "loss on ignition" (LOI) data, which show that unburned carbon in the flyash has returned to levels similar to pre-PRB days. The reduction in LOI has been aided by an increase in excess oxygen levels in the flue gas by about 1% over pre-PRB operating levels

## Ash control

The chemical composition of PRB ash poses a serious problem in boilers, and at Quindaro this was no exception. Ash now accumulates much more heavily on the boiler tube walls and in the boiler back pass. The existing Quindaro sootblowers were not totally effective in removing the ash, prompting plant management to install partial arc and selective pattern water lances,[4] manufactured by Diamond Power International.

These rotating lances spray a concentrated stream of water, at 300 psig, to those water-wall locations most prone to ash and slag buildups. The cyclone unit has two partial arc lances on the sidewalls and a selective pattern lance at the nose of the furnace cavity. Six lances are in the PC unit, with prime locations being the sidewalls and the wall opposite the burners. Selection of lance location is important as water lancing of furnace corners can cause thermal stress and shorten tube life.

# Other problems in burning PRB coal

When Unit 1 was converted to balanced draft operation in 1992, the induced draft (ID) fans that were added had sufficient air flow capacity to approach the maximum continuous rating (MCR) for the boiler. However, the new arrangement created several problems. One was a pressure versus flow condition which made control of the unit difficult and caused frequent unit trips. To overcome the problem, the ID fan motors were rewired to reduce their speed from 900 rpm to 720 rpm.

Unfortunately, the switch to PRB coal requires a 7.5% to 8.0% increase in fan capacity due to the coal's higher moisture content. This is being resolved by rewinding the motors back to their original 900 rpm design. ID fan capacity on Unit 2 is also a problem when operating at full load and burning PRB coal. Fan capacity modifications to Unit 2 are under design.

PRB coals contain significant quantities of calcium, which combines with the residual mineral oxides remaining after combustion. When exposed to moisture, the resulting substance becomes very hard and is difficult to remove.[5] To eliminate this problem and to reduce the frequency of off-line cleaning of the Ljungstrom air heater, sootblowers are being installed on the cold and hot air sides of the air heater.

# References

1 D. Merritt, "A Safe Approach to Power Plant Coal Fires," presentation at the ASME Research Committee on Power Plant & Environmental Chemistry fall committee meeting, September 17-20, 2000, Providence, Rhode Island.

2 J. Meinders, "Quindaro's Story: Conversion from Illinois Coal to Power River Basin Coal," presented at the Western & Advanced Fuels Conference 2000, August 6-8, Kansas City, Kansas

3 "ADA-249 Improves Cyclone Operations with PRB Coal," technical paper from ADA Environmental Services.

4 T. Gardner, "Water Cleaning Application Upgrades," presented at the Western & Advanced Fuels Conference 2000, August 6-8, Kansas City, Kansas.

5 B. Buecker, "The Air Heater: An Important but Sometimes Neglected Heat Exchanger," *Power Engineering*, Vol. 104, No.8, August 2000.

# Chapter 4

## Steim System Materials

4

## INTRODUCTION

Many different materials are used in the construction of a steam generating system. They range from mild carbon steel for condensate/feedwater piping and boiler waterwall tubes, to high-alloy steels for high-temperature superheater components, to copper alloys or other non-ferrous metals for many tube-in-shell heat exchangers.

This chapter outlines the most common materials that make up the steam-generating portion of power production or industrial plants. Certain groups of materials, such as the austenitic stainless steels, have members whose compositions are similar to but might vary slightly in one or two alloying elements. We will examine some of the basic properties of the materials and how alloying elements influence metal behavior.

## IRON AND STEEL

Humans have used iron for more than 4,500 years. Its chief benefits include strength, hardness, integrity at high temperatures, malleability when combined with other elements, and natural abundance. Iron has a very strong affinity for oxygen, which is why many natural iron deposits exist in combination with oxygen or oxygen-bearing compounds. This also explains why unprotected iron structures rust so easily. A little bit of moisture, and "voila!"—the material reacts with atmospheric oxygen and degrades into a pile of flakes. For many years, mankind's efforts have focused on production of iron-based alloys—the steels that resist corrosion but maintain the desired properties of strength and hardness.

The basic definition of steel is an alloy of iron and carbon. The addition of a wide variety of other elements to steel (principally metals but also some non-metals) imparts special properties. Before examining the steels used in boiler construction, it is important to examine the three most common molecular structures of metals and alloys. This is because alloying, heat treatment, and mechanical working often alter the molecular (or macromolecular) structure of a metal, which in turn influences such properties as strength, ductility, corrosion resistance, etc.

## Structures of metals

Metals form crystalline structures much like their non-metallic counterparts such as salt, limestone, and other minerals. This section provides a brief review of atomic structure and grain formation, which will be helpful when examining the properties of metals in a steam generator.

The three most common atomic structures of metals are body-centered cubic (BCC), face-centered cubic (FCC), and hexagonal close packed (HCP). These structures are shown in Figure 4-1. The common steel alloys for steam generating systems generally have either the BCC or FCC configuration. Alloy composition and temperature directly impact these configurations. This is illustrated in Figure 4-2, which shows the major effects of carbon content and temperature on the molecular structure of iron. This chart is known as the iron-carbon phase diagram. It illustrates the behavior of iron/steel over a wide range of temperatures and carbon content, but two features are important for this chapter. In the rectangular area

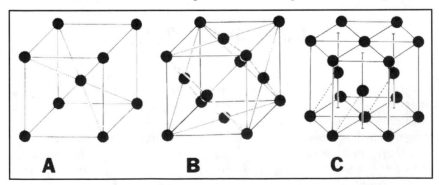

**Fig. 4-1:** The most common crystal structures of metals: A. body-centered cubic; B. face-centered cubic; C. hexagonal close packed (Reproduced with permission from *Corrosion Basics: An Introduction,* published by NACE International)

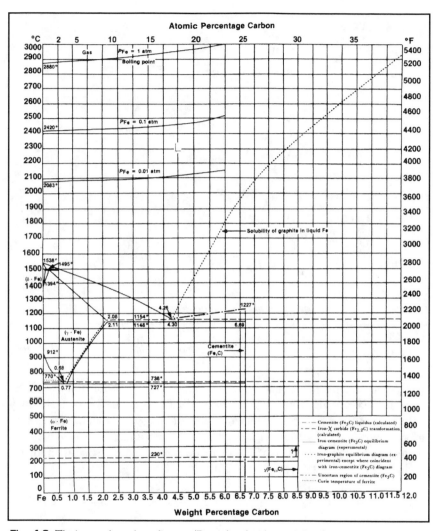

**Fig. 4-2:** The iron-carbon phase diagram (Reproduced with permission from *Corrosion Basics: An Introduction*, published by NACE International)

of the diagram between 0°C-727°C and 0%-6.7% carbon, the molecular structure exists in the BCC form, also known as the alpha (α) or ferrite structure. When carbon concentrations are below 0.025%, the carbon atoms sit midway between iron atoms along the edges of the atomic cube and at the center of the cube face. This is termed interstitial alignment. At carbon concentrations of 0.025%, iron becomes supersaturated with carbon, and the excess carbon combines with iron as cementite (Fe₃C), which appears as dark bands in the steel.

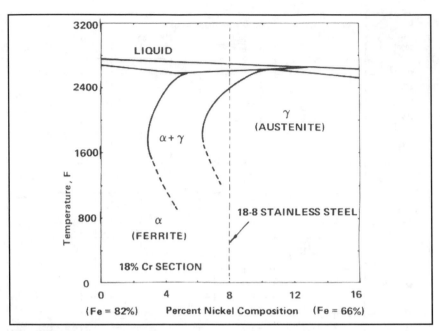

**Fig. 4-3:** Phase diagram for 18% chromium stainless steels, with variable nickel content and temperature (Reproduced with permission from *Corrosion Basics: An Introduction*, published by NACE International)

Notice the roughly triangular region midway up the left-hand side of the chart labeled γ - Fe. In this region, the molecular structure exists in the FCC arrangement, known as the austenite or gamma (γ) form. Carbon is more soluble in austenite due to its larger cell structure. The maximum carbon solubility in austenite is 2.1% at 1,148°C (2,098°F), which is much higher than carbon solubility in ferrite.

Figure 4-3 is the phase diagram for stainless steel containing 18% chromium, but with variable values of nickel concentration and temperature. At temperatures ambient to moderate (for boilers) and less than 8% nickel, the structure exists in the alpha configuration. As soon as the nickel content reaches 8%, the structure transforms to austenite regardless of temperature. This is a classic example of the influence of alloying elements on crystalline structure. Austenitic steels are used in many different applications.

Some metals alloyed with steel fit into the BCC or FCC lattices similarly to iron atoms. Defects are always present, such as those outlined in Figure 4-4. These irregularities include voids, interstitial alignment, crowding due to large atoms, and lattice disturbances. Movement of carbon atoms from interstitial points to other locations within the metal, such as grain boundaries, can lead to complications.

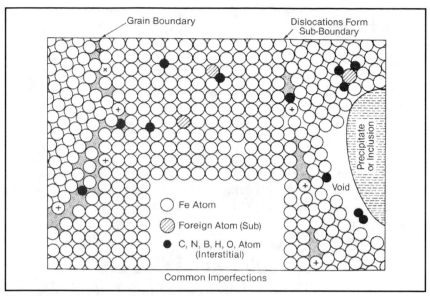

**Fig. 4-4:** Illustration of crystal defects and imperfections (Reproduced with permission from *Steam*, 40th ed., published by Babcock & Wilcox, a McDermott Company)

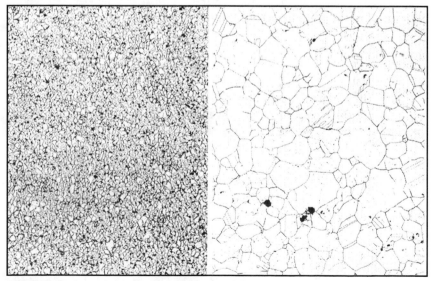

**Fig. 4-5:** Illustration of small and large grains (Source: Sinder, J.G., ed., *Combustion: Fossil Power*, Alstom)

## Grain formation

Consider a molten metal or alloy that is cooling to a solid. The crystalline structure that begins to emerge contains multiple FCC or BCC units. However, under normal conditions, the crystals only grow to a certain size, and then form a separate boundary with neighboring crystals. A metal normally does not consist of a single gigantic crystal, but rather is made up of many smaller crystals that fit together like a jigsaw puzzle (Fig. 4-5). These crystals are known as grains. The National Association of Corrosion Engineers has calculated that grain size typically ranges from 0.01 to 0.001 inches (0.25 to 0.025 mm). Grain formation is a very important concept, and has much bearing on the fabrication properties of a metal or alloy and its behavior in various environments.

# ALLOYS OF STEEL

Because steel is the most common material for boiler and steam generator construction, it is important to examine the principal alloying elements used in steel production and the special properties they provide.

## Carbon

In small percentages, carbon imparts strength and hardness to steel (Fig. 4-6). It gives steel the properties that allow it to be heat-treated and cold-worked to improve mechanical properties. The carbon concentration in steels commonly used for boiler materials ranges between 0.04% and 0.20%. Creep resistance is greater at carbon concentrations above 0.07%, but high carbon content, especially 2% or greater (the cast irons), increases the brittleness of steels. A classic example of brittle high-carbon steels is cast iron plumbing pipe, which can be separated with a chain cutter.

## Chromium

Chromium in low percentages improves the hardness and heat resistance of steels. A half percent of chromium also prevents graphitization, which is an important corrosion mechanism that is reviewed later.

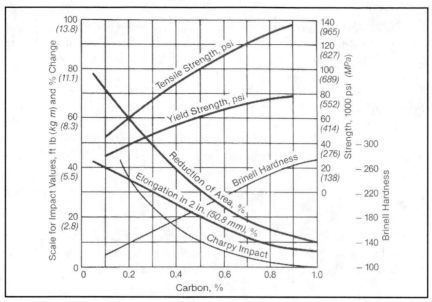

**Fig. 4-6:** Effects of carbon content on the properties of hot rolled steel (Reproduced with permission from *Steam*, 40th ed., published by Babcock & Wilcox, a McDermott Company)

At percentages above 12% by weight, chromium imparts corrosion resistance by causing the metal to form a surface layer of chromium oxide $(Cr_2O_3)$, which protects the alloy from its environment. This is the basis for the stainless steels.

## Molybdenum

Molybdenum increases strength, especially at high temperatures. It also enhances elasticity and other properties, including improved corrosion resistance in austenitic stainless steels. Low chromium, low molybdenum steels are the workhorse for boiler tube materials. Molybdenum, alloyed alone with steel or as the prime alloying agent, can be problematic. More details on this issue are discussed in the corrosion section of this chapter.

## Nickel

Nickel improves corrosion resistance and toughness. The extremely important austenitic stainless steels contain 18% chromium and 8% nickel, with minor quantities of additional alloys. Popular austenitic alloys include the 304, 316, and 317 series.

Nickel also serves as the base metal for a whole group of corrosion-resistant and/or high-temperature alloys. One common application of these materials is in flue gas desulfurization systems, where chemical environments are severe.

## Manganese

Manganese is added in small quantities to steel to react with residual sulfur, which otherwise increases brittleness.

## Silicon

Silicon in small quantities serves as a deoxidizer to react with any residual oxygen that might be in the steel. The basic grade of carbon steel contains 0.20% carbon, 0.45% manganese, and 0.25% silicon as the primary alloying agents.

## Phosphorous

Phosphorous in low concentrations serves as a hardening agent. Concentrations of 0.03% to 0.05% are common in steam generator materials.

This outlines the major alloying elements. Table 4-1 lists other minor alloying agents or steel additives and their primary functions. These minor elements are not used as regularly as those mentioned above.

# MECHANICAL AND CHEMICAL PROPERTIES OF STEEL

Steels and other metals are often defined or specified by their mechanical properties, which are of varying importance depending upon the application of the material. These properties include tensile strength, yield strength, hardness, and toughness. Figure 4-7 shows the relationship between tensile strength and yield strength. If increasing tension is applied to a material, it will stretch to a certain point and return to near its original dimensions if the stress is relieved. At the yield point, however, the material will continue to elongate with no further increase in tension. Failure is the eventual result. Hardness is the ability of the material to withstand deformation, and toughness is the ability to withstand localized stresses above the yield point.

| Element | Properties |
| --- | --- |
| Aluminum | Useful as a deoxidizer in small quantities. |
| Cobalt | Improves creep strength and solid solution strength. Will form carbides, thus preventing depletion of chromium by this process. |
| Copper | In small amounts improves resistance to atmospheric corrosion. |
| Niobium | Carbide former. |
| Nitrogen | Hardening agent in low alloy steels. Increases strength in stainless steels, but does not combine with chromium as carbon does. |
| Titantium | Deoxidizer and carbide former. |
| Tungsten | Improves creep strength and solid solution strength. |
| Vanadium | Strength, toughness, and hardening agent. Improves creep strength. Carbide former. |

**Table 4-1:** Minor alloying elements in steel (Reproduced with permission from *Steam*, 40th ed., published by Babcock & Wilcox, a McDermott Company)

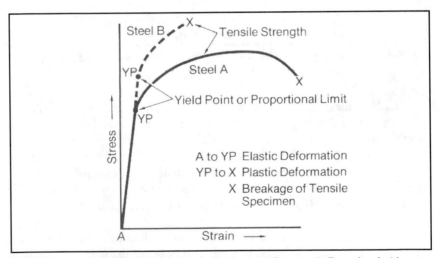

**Fig. 4-7:** Illustration of tensile strength and yield for two different steels (Reproduced with permission from *Combustion: Fossil Power*, Alstom)

Temperature, alloy content, and other factors all affect the properties listed above. Cooling techniques for molten steel, hot and cold working of solid steel, and post-cooling heat treatment are all important methods for establishing the desired properties of steels.

The technique used to cool molten steel has a great impact on the material structure. Figure 4-8 shows a time-temperature-transformation plot for a steel sample containing 0.8% carbon. If the steel is cooled slowly with the temperature remaining above 500°F (260°C), the atoms arrange themselves in the ferrite/iron carbide pattern shown earlier. However, if the steel is quenched with water, the rapid temperature drop will cause the formation of martensite, which has an entirely different crystal structure (body-centered tetragonal). Martensite is a very hard, but somewhat brittle material. The addition of other alloying elements will move the curves shown in the figure forward or backward, so cooling applications may have to be altered to obtain the desired property.

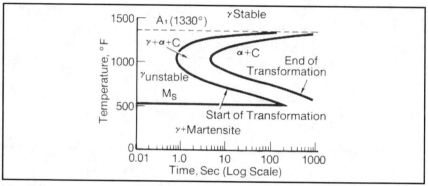

**Fig. 4-8:** Time-temperature-transformation plot for a 0.8% carbon steel (Reproduced with permission from *Combustion: Fossil Power*, Alstom)

"Cold working" (rolling, bending) is a very common technique for shaping metals into desired configurations and for improving properties. For example, cold working will strengthen the austenitic stainless steels, and Figure 4-9 shows the general effects of cold working on a metal. However, cold working has other effects that may sometimes be detrimental. Figure 4-10 illustrates cold-rolled iron samples. Note the elongated grains. Stresses in steel and other metals can cause the material to be susceptible to corrosion, particularly at grain boundaries.

A method commonly employed to relieve stress is heat treatment, wherein the metal, after fabrication, is heated to temperatures below the melting point from time periods of a few hours to a few days to allow the grains to realign. Stress relieving of cold-worked metals with heat treatment is known as annealing. Annealing is also used to improve ductility. Stress relieving is an important process following work such as welding on high-pressure boiler components, because welding often introduces localized stress points susceptible to corrosion. Specific procedures for heat-treating pressure parts under repair are outlined in the boiler codes of the American Society of Metallurgy Engineers (ASME). Stress relieving or tempering is employed on the martensitic stainless steels to improve strength and malleability. This allows martensitic steels to be used in more applications than would otherwise be possible.

From this overview of material properties, we will now examine the materials that actually make up a steam-generating unit.

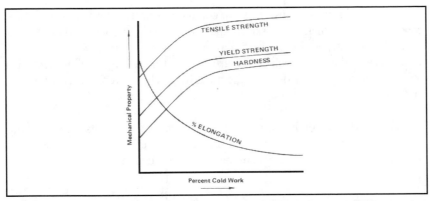

**Fig. 4-9:** Typical effect of cold working on the properties of a metal (Reproduced with permission from *Corrosion Basics: An Introduction*, published by NACE International)

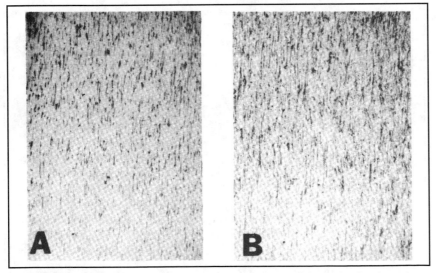

**Fig. 4-10:** Elongation of grains due to cold rolling: A. 30%, B. 60% (Reproduced with permission from *Corrosion Basics: An Introduction*, published by NACE International)

# COMMON STEELS USED FOR STEAM GENERATION COMPONENTS

Tables 4-2a and 4-2b list the most common steels used for steam generation construction and their chemical compositions. Generally, lower alloy content means lower cost, and this is reflected in the choice of materials. Waterwall tubes, which receive the most cooling, are usually constructed of low-alloy steels. For boiler designs of previous decades, the materials of choice were plain carbon steel and the 0.5 Mo carbon steels.

Note: When a number or numbers appear without any units it/they stand(s) for the common symbolism for steels in boiler and metallurgy books. The earlier example means "carbon steel containing 0.5% molybdenum."

| ALLOY | Product Form | ASME or ASTM Spec. | Grade | Minimum Tensile Strength ksi | Minimum Yield Strength ksi | Composition, % (a) | | | | | | | |
|---|---|---|---|---|---|---|---|---|---|---|---|---|---|
| | | | | | | C | Mn | P | S | Si | Ni | Cr | Mo |
| Carbon Steel Low-Strength | Tubes | SA-192 | ... | [47] | [26] | 0.06–0.18 | 0.27–0.63 | 0.048 | 0.058 | 0.25 | ... | ... | ... |
| | Tubes (ERW) | SA-178 | A | ... | ... | 0.06–0.18 | 0.27–0.63 | 0.050 | 0.060 | ... | ... | ... | ... |
| | Tubes (ERW) | SA-226 | ... | [47] | [26] | 0.06–0.18 | 0.27–0.68 | 0.050 | 0.060 | ... | ... | ... | ... |
| Intermediate Strength | Tubes | SA-210 | A-1 | 60 | 37 | 0.27 | 0.93 | 0.048 | 0.058 | 0.10 Min | ... | ... | ... |
| | Tubes (ERW) | SA-178 | C | 60 | 37 | 0.35 | 0.30 | 0.050 | 0.060 | ... | ... | ... | ... |
| | Pipe | SA-106 | B | 60 | 35 | 0.30 | 0.29–1.06 | 0.048 | 0.058 | 0.10 Min | ... | ... | ... |
| | Castings (b) | SA-216 | WCA | 60 | 30 | 0.25 | 0.70 | 0.040 | 0.045 | 0.60 | ... | ... | ... |
| | Structural Shapes | A36 | ... | 58 | 36 | 0.26 | ... | 0.040 | 0.05 | ... | | | |
| High Strength | Pipe | SA-106 | C | 70 | 40 | 0.35 | 0.29–1.06 | 0.048 | 0.058 | 0.10 Min | ... | ... | ... |
| | Plate | SA-299 | ... | 75 | 40 | 0.30 | 0.86–1.55 | 0.035 | 0.040 | 0.13–0.33 | ... | ... | ... |
| | Plate | SA-515 | 70 | 70 | 38 | 0.35 | 0.90 | 0.035 | 0.04 | 0.13–0.33 | ... | ... | ... |
| | Forging | SA-105 | ... | 70 | 36 | 0.35 | 0.60–1.05 | 0.040 | 0.050 | 0.35 | ... | ... | ... |
| | Casting (b) | SA-216 | WCB | 70 | 36 | 0.30 | 1.00 | 0.040 | 0.045 | 0.60 | ... | ... | ... |
| FERRITIC ALLOYS C–0.5 Mo | Tubes | SA-209 | T1 | 55 | 30 | 0.10–0.20 | 0.30–0.80 | 0.045 | 0.045 | 0.10–0.50 | ... | ... | 0.44–0.65 |
| 1 Cr–0.5 Mo | Forging | SA-336 | F12 | 70 | 40 | 0.10–0.20 | 0.30–0.80 | 0.040 | 0.040 | 0.10–0.60 | ... | 0.80–1.10 | 0.45–0.65 |
| | Tubes | SA-213 | T12 | 60 | 30 | 0.15 | 0.30–0.61 | 0.045 | 0.045 | 0.50 | ... | 0.80–1.25 | 0.44–0.65 |
| | Pipe | SA-335 | P12 | 60 | 30 | 0.15 | 0.30–0.61 | 0.045 | 0.045 | 0.50 | ... | 0.80–1.25 | 0.44–0.65 |
| | Plate | SA-387 | 12 Cl 2 | 65 | 40 | 0.17 | 0.36–0.69 | 0.035 | 0.040 | 0.13–0.32 | ... | 0.74–1.21 | 0.40–0.65 |
| | Forging | SA-182 | F12 | 70 | 40 | 0.10–0.20 | 0.30–0.80 | 0.040 | 0.040 | 0.10–0.60 | ... | 0.80–1.25 | 0.44–0.65 |
| 1.25 Cr–0.5 Mo | Tubes | SA-213 | T11 | 60 | 30 | 0.15 | 0.30–0.60 | 0.030 | 0.030 | 0.50–1.00 | ... | 1.00–1.50 | 0.44–0.65 |
| | Pipe | SA-335 | P11 | 60 | 30 | 0.15 | 0.30–0.60 | 0.030 | 0.030 | 0.50–1.00 | ... | 1.00–1.50 | 0.44–0.65 |
| | Plate | SA-387 | 11 Cl 2 | 75 | 45 | 0.17 | 0.36–0.69 | 0.035 | 0.040 | 0.44–0.86 | ... | 0.94–1.56 | 0.40–0.70 |
| | Forging | SA-182 | F11 | 70 | 40 | 0.10–0.20 | 0.30–0.80 | 0.040 | 0.040 | 0.50–1.00 | ... | 1.00–1.50 | 0.44–0.65 |
| | Casting (b) | SA-217 | WC6 | 70 | 40 | 0.20 | 0.50–0.80 | 0.040 | 0.045 | 0.60 | ... | 1.00–1.50 | 0.45–0.65 |

**Table 4-2a:** Common steam generator materials of construction (Reproduced with permission from *Combustion: Fossil Power*, Alstom)

| ALLOY | Product Form | ASME or ASTM Spec. | Grade | Minimum Tensile Strength ksi | Minimum Yield Strength ksi | Composition, % [a] | | | | | | | |
|---|---|---|---|---|---|---|---|---|---|---|---|---|---|
| | | | | | | C | Mn | P | S | Si | Ni | Cr | Mo |
| 2.25 Cr–1 Mo | Tubes | SA-213 | T22 | 60 | 30 | 0.15 | 0.30–0.60 | 0.030 | 0.030 | 0.50 | ... | 1.90–2.60 | 0.87–1.13 |
| | Pipe | SA-335 | P22 | 60 | 30 | 0.15 | 0.30–0.60 | 0.030 | 0.030 | 0.50 | ... | 1.90–2.60 | 0.87–1.13 |
| | Plate | SA-387 | 22 Cl 1 | 60 [c] | 30 [c] | 0.17 | 0.27–0.63 | 0.035 | 0.035 | 0.50 | ... | 1.88–2.62 | 0.85–1.15 |
| | | SA-387 | Cl 2 | 75 [d] | 45 [d] | | | | | | | | |
| | Forging | SA-182 | F22 | 75 | 45 | 0.15 | 0.30–0.60 | 0.040 | 0.040 | 0.50 | ... | 2.00–2.50 | 0.87–1.13 |
| | Casting [b] | SA-217 | WC9 | 70 | 40 | 0.18 | 0.40–0.70 | 0.040 | 0.045 | 0.60 | ... | 2.00–2.75 | 0.90–1.20 |
| 5 Cr–0.5 Mo | Tubes | SA-213 | T5 | 60 | 30 | 0.15 | 0.30–0.60 | 0.030 | 0.030 | 0.50 | ... | 4.00–6.00 | 0.45–0.65 |
| 9 Cr–1 Mo | Tubes | SA-213 | T9 | 60 | 30 | 0.15 | 0.30–0.60 | 0.030 | 0.030 | 0.25–1.00 | ... | 8.00–10.00 | 0.90–1.10 |
| AUSTENITIC STAINLESS ALLOYS 18 Cr–8 Ni | Tubes | SA-213 | TP 304H | 75 | 30 | 0.04–0.10 | 2.00 | 0.040 | 0.030 | 0.75 | 8.00–11.00 | 18.00–20.00 | ... |
| | Pipe | SA-376 | TP 304H | 75 | 30 | 0.04–0.10 | 2.00 | 0.040 | 0.030 | 0.75 | 8.00–11.00 | 18.00–20.00 | ... |
| | Plate | SA-240 | 304 | 75 | 30 | 0.08 | 2.00 | 0.045 | 0.035 | 1.00 | 8.00–10.50 | 18.00–20.00 | ... |
| | | SA-240 | 304H | 75 | 30 | 0.04–0.10 | 2.00 | 0.045 | 0.030 | 1.00 | 8.00–12.00 | 18.00–20.00 | ... |
| | Forging | SA-182 | F304H | 75 | 30 | 0.04–0.10 | 2.00 | 0.040 | 0.030 | 1.00 | 8.00–11.00 | 18.00–20.00 | ... |
| 18 Cr–10 Ni–Ti | Tubes [e] | SA-213 | TP 321H | 75 | 30 | 0.04–0.10 | 2.00 | 0.040 | 0.030 | 0.75 | 9.00–13.00 | 17.00–20.00 | ... |
| 18 Cr–10 Ni–Cb | Tubes [f] | SA-213 | TP 347H | 75 | 30 | 0.04–0.10 | 2.00 | 0.040 | 0.030 | 0.75 | 9.00–13.00 | 17.00–20.00 | ... |
| 16 Cr–12 Ni–2 Mo | Tubes | SA-213 | TP 316H | 75 | 30 | 0.04–0.10 | 2.00 | 0.040 | 0.030 | 0.75 | 11.00–14.00 | 16.00–18.00 | 2.00–3.00 |
| | Pipe | SA-376 | TP 316H | 75 | 30 | 0.04–0.10 | 2.00 | 0.040 | 0.030 | 0.75 | 11.00–14.00 | 16.00–18.00 | 2.00–3.00 |
| | Forging | SA-182 | F316H | 75 | 30 | 0.04–0.10 | 2.00 | 0.040 | 0.030 | 1.00 | 10.00–14.00 | 16.00–18.00 | |
| | Plate | SA-240 | 316H | 75 | 30 | 0.04–0.10 | 2.00 | 0.045 | 0.030 | 1.00 | 10.00–14.00 | 16.00–18.00 | 2.00–3.00 |
| | Structural Sheet | A167 | 316L | 70 | 25 | 0.03 | 2.00 | 0.045 | 0.03 | 1.00 | 10.00–14.00 | 16.00–18.00 | 2.00–3.00 |
| 25 Cr–12 Ni | Casting | SA-351 | CH20 | 70 | 30 | 0.20 | 1.50 | 0.040 | 0.040 | 2.00 | 12.00–15.00 | 22.00–26.00 | |

[a] Single values shown are maximums.
[b] Residual elements not to exceed 1.00%.
[c] Annealed.
[d] Normalized.
[e] Titanium content not less than four times carbon content and not more than 0.60%.
[f] Cb + Ta not less than eight times the carbon content and not more than 1.00%.

**Table 4-2b:** Common steam generator materials of construction (Reproduced with permission from *Combustion: Fossil Power*, Alstom)

However, these materials are subject to graphitization at weld seams, and a number of tube and pipe failures have occurred. A currently popular choice for waterwall and drum material is $\frac{1}{2}$ Cr – $\frac{1}{2}$ Mo steel ($\frac{1}{2}$% chromium and $\frac{1}{2}$% molybdenum), where the chrome inhibits graphitization. (A discussion of graphitization appears later in this chapter.)

Higher temperature zones, such as the superheater and reheater, often require more durable materials that can withstand the high temperatures and corrosive effects of flue gas byproducts. As Table 4-3 indicates, a percentage or two increase in chromium and smaller increases in molybdenum improve temperature resistance. The 1 Cr – $\frac{1}{2}$ Mo steels are becoming popular for intermediate temperature functions, and the $2\frac{1}{4}$ Cr – 1 Mo steels, particularly SA-213-T22, still find much application. For high-temperature service, stainless steels of the 304, 310, and 347 variety are common, although Babcock & Wilcox reports that the 9 Cr – 1 Mo – V ferritic steel is a popular replacement for these materials in new units. For non-pressure, high-temperature components the common choices are the 25 Cr – 12 Ni and 25 Cr – 20 Ni steels. These materials maintain structural strength up to 1,400°F (760°C) to 1,500°F (816°C).

| Specification | Nominal Composition | Product Form | Min Tensile, ksi | Min Yield, ksi | High Heat Input Furn Walls | Other Furn Walls and Enclosures | SH RH Econ | Unheated Conn Pipe < 10.75 in. OD | Headers and Pipe > 10.75 in. OD | Drums | Recomm Max Use Temp, F | Notes |
|---|---|---|---|---|---|---|---|---|---|---|---|---|
| SA-178A | C-Steel | ERW tube | (47.0) | (26.0) | X | X | X | | | | 950 | 1,2 |
| SA-192 | C-Steel | Seamless tube | (47.0) | (26.0) | X | X | X | X | | | 950 | 1 |
| SA-178C | C-Steel | ERW tube | 60.0 | 37.0 | | X | X | | | | 950 | 2 |
| SA-210A1 | C-Steel | Seamless tube | 60.0 | 37.0 | X | X | X | X | | | 950 | |
| SA-106B | C-Steel | Seamless pipe | 60.0 | 35.0 | | | | X | X | | 950 | 3 |
| SA-178D | C-Steel | ERW tube | 70.0 | 40.0 | X | X | X | | | | 950 | 2 |
| SA-210C | C-Steel | Seamless tube | 70.0 | 40.0 | | X | X | X | | | 950 | |
| SA-106C | C-Steel | Seamless pipe | 70.0 | 40.0 | | | | X | X | | 950 | 3 |
| SA-216WCB | C-Steel | Casting | 70.0 | 36.0 | | X | X | X | X | | 950 | |
| SA-105 | C-Steel | Forging | 70.0 | 36.0 | | X | X | X | X | | 950 | 3 |
| SA-181-70 | C-Steel | Forging | 70.0 | 36.0 | | X | X | X | X | | 950 | 3 |
| SA-266Cl2 | C-Steel | Forging | 70.0 | 36.0 | | | | X | | | 800 | |
| SA-516-70 | C-Steel | Plate | 70.0 | 38.0 | | | | X | X | | 800 | |
| SA-266Cl3 | C-Steel | Forging | 75.0 | 37.5 | | | | X | | X | 800 | |
| SA-299 | C-Steel | Plate | 75.0 | 40.0 | | | | X | | X | 800 | |
| SA-250T1a | C-Mo | ERW tube | 60.0 | 32.0 | | X | X | | | | 975 | 4,5 |
| SA-209T1a | C-Mo | Seamless tube | 60.0 | 32.0 | | X | X | X | | | 975 | 4 |
| SA-335P1 | C-Mo | Seamless pipe | 55.0 | 30.0 | | | | X | | | 875 | |
| SA-250T2 | 1/2Cr-1/2Mo | ERW tube | 60.0 | 30.0 | X | | X | | X | | 1025 | 6,7 |
| SA-213T2 | 1/2Cr-1/2Mo | Seamless tube | 60.0 | 30.0 | X | | X | | | | 1025 | 6 |
| SA-250T12 | 1Cr-1/2Mo | ERW tube | 60.0 | 32.0 | | | X | | | | 1050 | 5,7 |
| SA-213T12 | 1Cr-1/2Mo | Seamless tube | 60.0 | 32.0 | | | X | | | | 1050 | 5,7 |
| SA-335P12 | 1/2Cr-1/2Mo | Seamless pipe | 60.0 | 32.0 | | | X | | X | | 1050 | 8 |
| SA-250T11 | 1-1/4Cr-1/2Mo-Si | ERW tube | 60.0 | 30.0 | | | X | | | | 1050 | 8 |
| SA-213T11 | 1-1/4Cr-1/2Mo-Si | Seamless tube | 60.0 | 30.0 | | | X | | | | 1050 | 5 |
| SA-335P11 | 1-1/4Cr-1/2Mo-Si | Seamless pipe | 60.0 | 30.0 | | | | X | X | | 1050 | |
| SA-217WC6 | 1-1/4Cr-1/2Mo | Casting | 70.0 | 40.0 | X | X | X | X | X | | 1100 | |
| SA-250T22 | 2-1/4Cr-1Mo | ERW tube | 60.0 | 30.0 | | | X | | | | 1115 | 5,7 |
| SA-213T22 | 2-1/4Cr-1Mo | Seamless tube | 60.0 | 30.0 | | | X | | | | 1115 | |
| SA-335P22 | 2-1/4Cr-1Mo | Seamless pipe | 60.0 | 30.0 | | | | X | X | | 1100 | |
| SA-217WC9 | 2-1/4Cr-1Mo | Casting | 70.0 | 40.0 | | | X | X | X | | 1115 | |
| SA-182F22Cl1 | 2-1/4Cr-1Mo | Forging | 60.0 | 30.0 | | | X | | X | | 1115 | |
| SA-336F22Cl1 | 2-1/4Cr-1Mo | Forging | 60.0 | 30.0 | | | | | X | | 1100 | |
| SA-213T91 | 9Cr-1Mo-V | Seamless tube | 85.0 | 60.0 | | | X | | | | 1200 | |
| SA-335P91 | 9Cr-1Mo-V | Seamless pipe | 85.0 | 60.0 | | | | X | X | | 1200 | |
| SA-182F91 | 9Cr-1Mo-V | Forging | 85.0 | 60.0 | | | X | | | | 1200 | |
| SA-336F91 | 9Cr-1Mo-V | Forging | 85.0 | 60.0 | | | | | X | | 1200 | |
| SA-213TP304H | 18Cr-8Ni | Seamless tube | 75.0 | 30.0 | | | X | | | | 1400 | |
| SA-213TP347H | 18Cr-10Ni-Cb | Seamless tube | 75.0 | 30.0 | | | X | | | | 1400 | |
| SA-213TP310H | 25Cr-20Ni | Seamless tube | 75.0 | 30.0 | | | X | | | | 1500 | |
| SB-407-800H | Ni-Cr-Fe | Seamless tube | 65.0 | 25.0 | | | X | | | | 1500 | |
| SB-423-825 | Ni-Fe-Cr-Mo-Cu | Seamless tube | 85.0 | 35.0 | | | X | | | | 1000 | |

Notes:
1. Values in parentheses are not required minimums, but are expected minimums.
2. Requires special inspection if used at 100% efficiency above 850F.
3. Limited to 800F maximum for piping 10.75 in. OD and larger and outside the boiler setting.
4. Limited to 875F maximum for applications outside the boiler setting.
5. Requires special inspection if used at 100% efficiency.
6. Maximum OD temperature is 1025F. Maximum mean metal temperature for Code calculations is 1000F.
7. Requires use of a Code Case now. Will not later.
8. 32 ksi minimum yield requires use of Code Case 2070, which is being incorporated into the Code.

**Table 4-3:** Common steam generator materials of construction (Reproduced with permission from *Steam*, 40th ed., published by Babcock & Wilcox, a McDermott Company)

# CORROSION AND FAILURE MECHANISMS OF BOILER MATERIALS

All of the components on the combustion side of a steam-generating unit are subject to harsh and potentially corrosive conditions. Failures may occur through mechanical or chemical methods.

## Mechanical failure mechanisms

A phenomenon of great importance in high-temperature applications is creep. Creep has several meanings, but in metallurgical terms it is defined as the deformation of material that is subjected to continual stress. Boiler tubes, supports, and other components are designed to withstand creep, and when equipment is kept clean and in good condition, creep is usually not serious. Obviously, as temperature increases, metal strength decreases. This loss of strength is tolerable within a metal's normal working temperature, but if temperatures exceed this range, the metal will begin to deform. Creep may be initiated and propagated by internal tube deposition, which restricts heat transfer from the combustion zone to the cooling medium whether it be water or steam. The deposition would be iron oxide corrosion products, mineral salts from condenser leaks, etc. Material strength rapidly falls off as the temperature climbs above the maximum working temperature (Fig. 4-11).

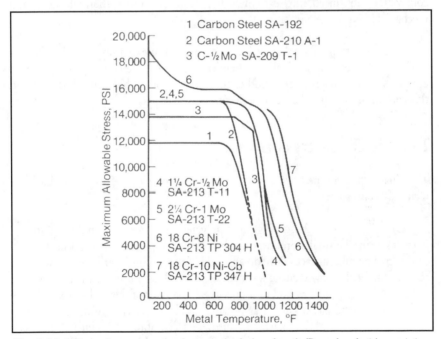

**Fig. 4-11:** Effects of temperature on the strength of selected steels (Reproduced with permission from *Combustion: Fossil Power*, published by Alstom)

Grain size may also influence creep. Quoting from *Steam*:

*At lower temperatures, a steel with small grains (fine grain size) may be stronger than the same steel with fewer large grains (coarse grain size) because the grain boundaries act as barriers to slip. At higher temperatures, where thermally activated deformation can occur, a fine grain structure material may be weaker because the irregular structure at the grain boundaries promotes local creep.*

High temperature steel fabrication techniques may include processes to enlarge grain size.

Another failure mechanism is known as fatigue. Have you ever broken a wire by repeatedly bending it back and forth at one point? This is fatigue. During boiler start-ups and shutdowns, the boiler metal expands and shrinks. Over time, these cycles can cause failures in such locations as boiler buckstays and membrane webs. Fatigue is a common failure mechanism at tube/tube support interfaces. Instead of the tube support failing, a crack initiates and propagates in the tube.

Water lancing is sometimes used to remove difficult ash deposits. However, cold water introduced into extremely hot tubes can cause thermal shock and mechanical failures.

Chain reaction failures can be induced by the failure of a single tube. Whether the initial failure is mechanically or chemically induced, the failure allows steam to escape as a jet. This may cut another tube, and so on. The author has observed this condition on more than one occasion.

## Fireside chemistry corrosion

Chapter 3 examined ash corrosion. What are some of the other "fireside" corrosion mechanisms?

One phenomenon becoming more prevalent is corrosion in reducing atmospheres. During "normal" combustion, when excess air persists throughout the boiler, the fireside tube surfaces develop an oxide coating. This layer is generally rather stable. One of the new techniques for $NO_x$ control is to fire the main burners at sub-stoichiometric airflow and introduce the remainder of the air higher up in the boiler. This technique is known as overfire air. The fuel-rich firing in the main burners allows most of the fuel-bound nitrogen to combine as $N_2$ rather than react with oxygen to form nitrogen oxides. Unfortunately, this also creates a reducing atmosphere in the zone between the main burners and the overfire air ports. The lean atmosphere allows some of the sulfur in the coal, which would normally be converted to $SO_2$, to remain in a sulfide state. The sulfide reacts with the tube metal to form a layer of iron sulfide, which is not protective and can spall off the tubes.

The reducing atmosphere before the overfire air ports contains elevated levels of carbon monoxide (CO). Carbon monoxide may react with steel to allow carbon to penetrate into the metal:

$$2CO \rightarrow C \text{ (solid)} + CO_2 \uparrow$$

This can cause graphitization, which weakens the steel. One possible solution to this problem is to design these regions with steels containing carbon of similar concentration to the projected environment. Like equilibrium reactions of chemicals, the carbon in the steel inhibits diffusion of carbon from the boiler gases.

Graphitization is a phenomenon that could be called a mechanically-influenced chemical phenomenon. If carbon steel becomes overheated, carbon atoms tend to migrate towards grain boundaries. They will combine to form graphite nodules, which in severe cases line up in a chain-like pattern. This influences the mechanical properties and strength of the steel. One situation in which this is particularly critical is welding. Highly concentrated temperatures at the welding zone may induce graphitization in a localized area. Many pipe and tube failures have been attributed to metallurgical defects initiated by improper welding techniques.

One reasonably successful solution to graphitization is the addition of chromium to carbon steels. The chromium reacts with carbon to form chromium carbide ($Cr_3C$), thereby preventing the formation of graphite nodules. This solution must be viewed with caution, however, as the reaction of chromium and carbon removes chromium from grain boundaries and lowers the ability of the metal to form a protective oxide coating. Corrosion has been known to start at chromium-depleted sites. Some of the minor elements outlined in Table 4-1 will preferentially react with carbon, which allows the chromium to remain undisturbed.

Welding presents another potential problem, especially where two different metals are joined. Consider the case of a low-alloy carbon steel welded to a higher-alloy steel or even stainless steel. This is most common in superheaters. The two metals have different coefficients of expansion, which places stress upon the weld during temperature changes. Fatigue may eventually cause weld fracture.

Materials used in conventional boilers are also well suited for use in heat recovery steam generators (HRSG). The combination of lower heat fluxes and natural gas firing subject HRSG components to less harsh conditions than those of coal-fired boilers. One situation of concern in conventional boilers and HRSGs is flow-assisted-corrosion (FAC). This waterside phenomenon strips the protective oxide layer from carbon-steel pipe surfaces in systems treated with oxygen scavengers. FAC is particularly troublesome at elbows in feedwater lines, economizers, and in sharp radius fittings in HRSGs. In the last decade, personnel at several power plants in the U.S. have been killed by FAC-induced pipe failures. Sharp bends are common with HRSG steam generator tubes, and FAC has already become a problem in a number of these systems, even though they are not very old. The problem can be solved in the design stage by fabricating elbows and other flow change fittings out of the low-chromium, low-molybdenum steels.

# MATERIALS FOR OTHER STEAM GENERATING COMPONENTS

Although this book focuses on boilers and the actual steam generating process, material selection is also important for other system components.

## Condensate/feedwater system

Condensate and feedwater piping is always constructed of mild carbon steel, although the low-chromium, low-molybdenum alloys are finding use in areas prone to FAC.

The primary objective in the condensate/feedwater system is heat transfer. Efficient heat exchange is important to bring the feedwater temperature up to design levels for introduction to the boiler. Table 4-4 illustrates common materials for condenser and heat exchanger tubing and thermal conductivity of each. Heat exchange issues are why many condensers and heat exchanger tubes were fabricated from copper alloys, especially in units constructed in the 1960s and 1970s. Copper is an excellent heat transfer material.

A very popular copper alloy for heat exchangers installed during this time period was Admiralty metal, which contains 70% copper, 29% zinc, and usually 1% tin or phosphorous as a stabilizing agent. Subsequent history has shown that this

| Material | Thermal Conductivity BTU/(ft$^2$-hr-°F-in) |
|---|---|
| Admiralty Metal | 768 |
| 90-10 Copper-Nickel | 312 |
| 70-30 Copper-Nickel | 252 |
| 304 & 316 Stainless Steel | 113 |
| Titanium | 114 |
| Carbon Steel | 360 |

**Table 4-4:** Common heat exchanger tube materials and their heat transer coefficients (Source: "*Properties of Some Metals and Alloys*", distributed by the Nickel Development Institute)

material is susceptible to chemical and mechanical corrosion, which may shorten tube life and, more importantly, introduce copper ions to boiler water. Copper deposits in boilers generate corrosion cells. In units operating at 2,400 psig and above, copper will carry over with steam and then redeposit on high-pressure turbine blades. Just a few pounds of copper deposition can reduce turbine capacity by several percent. The principal steam-side corrodent of Admiralty is the combination of ammonia and oxygen.

For these reasons, plant designers or plant personnel who have to retube systems favor more durable materials. These include the 90-10 and 70-30 copper-nickel alloys and stainless steel. Even the copper-nickels are falling into disfavor, however, and stainless steel is becoming the material of choice for freshwater applications. Seawater presents a special concern due to its high chloride and dissolved solids content. A popular material for seawater service is titanium.

Materials for feedwater heater tubes include carbon steel, Admiralty metal, the copper-nickels, and stainless steel. Carbon steel is not overly popular anymore, and Admiralty is subject to similar corrosion in feedwater heaters as in condensers. Admiralty, the copper-nickels, and carbon steel may suffer severe corrosion during outages if air enters the feedwater system. The consequences include exfoliation and selective leaching of the material. The preferred material for feedwater heater tubing has become stainless steel.

## Steam turbines

Turbines are complex pieces of machinery that must function with precision while being subjected to high stress. Inlet temperatures to modern turbines usually are at or near 1,000°F (538°C), although some very high-pressure units may operate with inlet turbine temperatures at 1,050°F (566°C).

Common materials for these high-temperature, high-stress regions are the martensitic stainless steels, which contain 12% chromium and 1% (or so) of manganese and silicon. At the low-pressure end of the turbine, where condensate begins to form, different materials function better. Stainless steel was once the dominant material for low-pressure turbine blades, but titanium is becoming more popular because of its better corrosion resistance.

## Flue gas desulfurization systems

Liquids and gases in flue gas desulfurization (FGD) systems may be some of the most corrosive fluids within the entire plant. Most FGD systems are either of the wet-limestone or lime variety, in which the incoming flue gas is quenched by the alkaline limestone or lime slurry, which proceeds to react with sulfur dioxide ($SO_2$). A variety of corrosive environments may exist within the scrubbing vessel.

The initial contact point of the flue gas with the slurry (known as wet/dry inter-face) is a site of localized acidic conditions, where the pH might be in the one to two range. The reaction products of the alkaline slurry and $SO_2$ are calcium sulfite hemihydrate $(CaSO_3 \cdot \frac{1}{2} H_2O)$ and, in forced-oxidation systems, gypsum $(CaSO_4 \cdot 2H_2O)$. They deposit on tank and wet/dry interface duct walls and estab-lish sites for underdeposit corrosion. Flue gas outlet ducts may develop pockets of acidic moisture, especially in FGD systems with bypass flue gas reheat. Many FGD systems operate in what is commonly called "closed loop" mode to minimize discharge of liquids. This concentrates reaction products—most notably, chlorides and to a lesser extent, fluorides. Chloride concentrations are often in the tens-of-thousands of part-per-million (ppm) range, and have been know to exceed 100,000 ppm.

The history of FGD technology has shown a progression of materials to pre-vent corrosion. Materials that have been tried (but have not always held up) in severe FGD environments include the 304L, 316L, 317LM, and 904L stainless steels. Coatings and rubber linings have often been just as difficult to maintain. Even titanium will fail under certain conditions.

During the author's utility career, the inlet ducts to two FGD absorber tow-ers suffered extensive underdeposit corrosion. The original material was 904L stainless steel. The deposits primarily consisted of sludge-like $CaSO_3 \cdot \frac{1}{2} H_2O$, which allowed penetration of high-chloride liquid. Plant management had the walls clad with titanium sheets, but within a couple of years the titanium had also failed due to underdeposit corrosion.

In the 1980s, researchers began to discover that several of the nickel-based alloys were resistant to low-pH, high-chloride corrosion. Table 4-5 outlines these materials. Because these alloys are much more expensive than the common stainless steels, plant personnel at existing facilities adopted the technique of "wall-papering," whereby thin sheets of the material are welded or explosively clad to the existing metal to prevent corrosion. New construction now uses clad steel plates as the original material.

| Material | Ni(%) | Cr(%) | Mo(%) | Other |
|----------|-------|-------|-------|-------|
| C-276 | 56 | 16 | 16 | 4% Tungsten |
| C-22 | 58 | 22 | 13 | 3% Tungsten |
| Alloy 59 | 59 | 23 | 16 | |
| Alloy 625 | 61 | 22 | 9 | 4% Niobium |

**Table 4-5:** Composition of corrosion resistant FGD Alloys (Ross, R.W. Jr., *The Evolution of Stainless Steel and Nickel Alloys in FGD Materials Technology*, Paper presented at the EPRI/EPA/DOE 1993 $SO_2$ Control Symposium, Boston, MA, 1993)

# Information note

Readers who wish to learn more about steam generator materials may consult the references at the end of this book. Both Babcock & Wilcox's and Alstom's books go into considerable detail. The National Association of Corrosion Engineers (NACE) is a valuable resource. The Nickel Development Institute (NiDI) publishes or distributes a great deal of information on materials and materials performance. Most of the NiDI information can be obtained free or at nominal cost.

# Chapter 5

## Air Pollution Control

Control of air pollution—it's an issue of tremendous importance to power plant owners past, present, and future. While the industry focus for years has been on nitrogen oxides ($NO_x$), sulfur dioxide ($SO_2$), and particulate emissions, other pollutants are receiving increased scrutiny. Plant owners face certain or potential regulations for mercury, fine particulate matter (particles less than 2.5 microns in diameter), and carbon dioxide ($CO_2$).

Earlier chapters outlined several boiler types that inherently limit pollutant emissions. These include fluidized-bed and coal gasification steam generators. This chapter examines other technologies, including traditional and cutting-edge methods that are most popular and promising for air pollutant control. In many cases, combinations of methods may be necessary, or most economical, to bring specific pollutant emissions within established or future guidelines. The United States Department of Energy (DOE) has sponsored many important pollutant control projects through their Clean Coal Technology program, and illustrations of some of them appear in this chapter.

Power generation from fossil fuels other than coal, particularly natural gas, is much simpler with regard to pollution control, with the exception of $NO_x$. One of the advantages that is touted about natural gas-fired units is that they produce about half the $CO_2$ as coal-fired plants. This issue will also be covered.

# AIR QUALITY AND POLLUTANT DISCHARGE REGULATIONS

Air pollution control laws have been on the books for almost 40 years. The actual laws as written by Congress are quite complicated; a condensed compilation of the guidelines can be found in *Clean Air Act Permitting: A Guidance Manual* (PennWell, 1995). This section outlines important details from the laws, especially with regard to emissions limits that are in place or anticipated in the future.

The first national air pollution legislation in the U.S., known simply as the Clean Air Act, was enacted in 1963. This initial legislation was designed to guide the states in dealing with air pollution control issues. Additional legislation was proposed and passed later in the 1960s, but the real turning point came with the establishment of the Environmental Protection Agency (EPA) in 1970 and the subsequent passage of the Clean Air Act Amendments (CAAA) in December of that year. Congress and the EPA modified the CAAA in 1977 and again in 1990.

From the outset, the EPA was tasked with developing National Ambient Air Quality Standards (NAAQS), which with passage of the 1990 CAAA, evolved into the stipulation that "Each state is responsible for meeting the NAAQS for the six criteria air pollutants (nitrogen oxide, sulfur dioxide, carbon monoxide, ozone, particulate matter and lead). . . ." The EPA has since developed NAAQS guidelines for fine particulates less than 2.5 microns in diameter (PM2.5), although this proposal has been the subject of much debate. Utility and industrial owners have argued that fine particulate control would be too expensive and not worth the extra costs, while environmental leaders claim the laws will significantly improve public health.

In February 2001, the U.S. Supreme Court issued a ruling that allows the EPA to establish PM2.5 and ozone guidelines. At the time of this writing, the EPA has not established limits on PM2.5 emissions from power plants.

Another issue affecting utilities is regional haze regulations, which are intended to improve visible air quality, especially in pristine locations such as areas bordering national parks. Pristine locations are known as Class 1 areas, and

*...states will need to prepare strategies and revisions . . . to regulations to reduce and eventually eliminate existing visibility impairment in and near Class I areas. The first long-term strategy will cover 10 to 15 years with reassessment and possible revisions to the strategy in 2018 and every 10 years thereafter. Ultimately, the goal is to reach natural background conditions in 60 years. . . . So, if the EPA finalizes the RH rule, states must identify plants as early as 2004. Controls representing "the best available technology" must be then installed within five years. Subsequent revisions are required in 2018 and every 10 years thereafter. . . . several larger coal fired system owners believe the RH rule will force FGD (or similar [technologies] at several plants during the period from 2011-2015).*

Appendix 5-1 provides a good summary of air pollution control legislation as of 2000, including tables of the NAAQS and emissions regulations. The reader is encouraged to examine this appendix before proceeding.

| |
|---|
| Title I Nonattainment Requirements |
| Title III Air Toxics |
| Title IV Acid Deposition Control |
| Title V Operating Permits |
| Title VII Enforcement |

**Table 5-1:** List of Clean Air Act Amendment Titles

The 1990 CAAA were separated into several sections known as Titles (Table 5-1). Title IV of the CAAA has been of particular importance to the electric utility industry, as this is the section dealing with "acid rain" and its $SO_2$ and $NO_x$ precursors. With regard to $SO_2$, the primary stipulation of Title IV called for a reduction of $SO_2$ emissions from 20 million tons per year to 10 million tons annually. The $SO_2$ regulations have been implemented in two phases. In Phase I, the EPA identified the 261 "top emitters" and required $SO_2$ reduction programs to be in place by January 1, 1995. Phase II required reductions at the 2,500 remaining boilers January 1, 2000.

*In Phase II, all existing boilers must meet $SO_2$ emission levels of 1.2 lb/$10^6$ Btu [0.52 kg/$10^6$ kJ] and a sliding-scale percent reduction of 70 to 90 percent, depending upon the input sulfur content. The resultant $SO_2$ emission levels are generally 0.3 lb/$10^6$ Btu [0.13 kg/$10^6$ kJ] for low sulfur-coals and 0.6 lb/$10^6$ Btu [0.26 kg/$10^6$ kJ] for high-sulfur coals. Moreover, the CAAA calls for $SO_2$ emissions to be limited to 9.48 million tons [8.60 metric tons] per year between 2000 and 2009 and 8.95 million tons per year [8.12 metric tons] thereafter...*

Fuel switching was instituted as a quick fix to reduce $SO_2$ emissions at many utilities, but this technique will probably not completely suffice for the guidelines that are anticipated. As a consequence, the scrubber market appears ready to blossom as the CAAA is fully implemented.

Appendix 5-1 details $SO_2$ emissions limits for various categories of fossil-fired boilers. The limits are broken down according to type of fossil fuel and heat input. The largest boilers must meet the most stringent guidelines.

$NO_x$ control was also set up in two phases, as determined by boiler type. This is illustrated in Figure 5-1, which shows original $NO_x$ limits for common boiler categories. The Phase I compliance date was January 1, 1996, while the Phase II compliance date is set at January 1, 2004.

| Group 1 Boiler Type | Group 2 Boiler Type | Phase I $NO_x$ Emission Limits[a] (lb/10⁶Btu) | Phase II $NO_x$ Emission Limits[a] (lb/10⁶ Btu) |
| --- | --- | --- | --- |
| Tangentially fired boilers | | 0.45 | 0.40 |
| Dry-bottom wall-fired boilers[b] | | 0.50 | 0.46 |
| | Cell-burner boilers | | 0.68 |
| | Cyclone boiler >155 MWe | | 0.86 |
| | Wet-bottom wall-fired boilers >65 MWe | | 0.84 |
| | Vertically fired boilers | | 0.80 |

[a]Emission limits are lb/10⁶Btu of heat input on an annual average basis.

[b]Other than units applying cell-burner technology.

**Fig. 5-1:** CAAA $NO_x$ emission limits (Source: *Clean Coal Technology Demonstration Program: Program Update 2000*; U.S. Department of Energy, Washington, D.C., April 2001)

Unfortunately for the environment and for the utility managers trying to control pollution, "acid rain" is not the only problem caused by $NO_x$ discharges. Nitrogen oxides participate in photochemical reactions with hydrocarbons to produce smog and ground-level ozone. Many areas of the country, and especially the eastern states, fall into what are known as ozone non-attainment areas. This complicating factor will drive $NO_x$ emissions limits downward, and the EPA has proposed more stringent $NO_x$ limits in 22 eastern states and the District of Columbia to reduce ozone concentrations. The projected discharge limit at the time of this writing is 0.15 pounds per 10⁶ Btu (0.064 kg/10⁶ kJ), equivalent to 123 parts-per-million by volume (ppmv) at 3% $O_2$ in the stack. New coal plants, as regulated by federal New Source Performance Standards, face $NO_x$ emissions limits of 1.6 lb per megawatt-hour (0.73 kg/MWhr). This corresponds closely to the 0.15 lb/$10^6$ Btu limit mentioned earlier.

Regulations for gas turbine $NO_x$ emissions are even tighter, and in some areas the limits are as low as 2 ppmv. Natural gas-fired plants inherently produce lower $NO_x$ than coal-fired boilers (as outlined in the $NO_x$ control section), and so environmental regulators are comfortable with stringent emissions limits.

Issues that many analysts believe will impose near-term difficulties on the power industry are those of PM2.5 and mercury discharges. Regulations pertaining to PM2.5 and mercury may also affect previously established $SO_2$ and $NO_x$ guidelines, as both are known to generate very fine particulates and mists.

*The most effective options for controlling PM2.5 emissions are more stringent limits on sulfur oxides ($SO_x$), nitrogen oxides ($NO_x$) and sub-micron particulate emissions from fossil boilers. . . . Data on fine particulate matter shows sulfates and nitrates are the most abundant and potentially damaging species in atmospheric aerosols. Sulfates comprise as much as half of fine particulate matter mass in both rural and urban areas of the United States.*

As a following section will illustrate, backend technologies are already in place for control of fine particulates. As a result, what effect new PM2.5 regulations will have on $SO_2$ and $NO_x$ limits is still the subject of conjecture. Mercury control looms on the horizon.

*The median percentage mercury concentration in U.S. coals ranges from about 0.03 parts-per-million by weight (ppmw) to 0.24 ppmw. With more than 800 million tons [726,000,000 metric tons] of coal being burned by utilities in the United States this source has potential emissions of more than 100 tons [90.7 metric tons] of mercury each year, which pegs utilities as the single largest source of utility emissions.*

The EPA is expected to mandate mercury emissions reductions of up to 90%. Control methods are the subject of considerable research within the electric utility industry.

What can utility managers do to meet present and future guidelines? The following sections examine the primary control methods for the pollutants outlined above, and examine combinations of technologies that may be required to meet increasingly stringent regulations. The chapter will conclude with a look at the potential issues of carbon dioxide control and that of hazardous air pollutants.

# SULFUR DIOXIDE CONTROL

As chapter 3 outlined, the amount and makeup of impurities differs widely between coals and are a function of many variables including coal age, the environment in which the original plants developed and thrived, and minerals transported from surrounding and overlying rock formations and soil. Of particular importance are sulfur and nitrogen content. Particulate emissions are in large part influenced by the combustion process, although the ash chemical makeup can significantly influence design and operation of particulate removal systems. Tables 5-2 and 5-3 offer a good range of sulfur content in many of the major coals in the U.S. Notice the very low sulfur content of the Wyoming and Montana coals as compared to that of Illinois, Pennsylvania, and other eastern coals.

| | Coal Rank | | | | Coal Analysis, Bed Moisture Basis | | | | | | Rank | Rank |
|---|---|---|---|---|---|---|---|---|---|---|---|---|
| No. | Class | Group | State | County | M | VM | FC | A | S | Btu | FC | Btu |
| 1 | I | 1 | Pa. | Schuylkill | 4.5 | 1.7 | 84.1 | 9.7 | 0.77 | 12,745 | 99.2 | 14,280 |
| 2 | I | 2 | Pa. | Lackawanna | 2.5 | 6.2 | 79.4 | 11.9 | 0.60 | 12,925 | 94.1 | 14,880 |
| 3 | I | 3 | Va. | Montgomery | 2.0 | 10.6 | 67.2 | 20.2 | 0.62 | 11,925 | 88.7 | 15,340 |
| 4 | II | 1 | W.Va. | McDowell | 1.0 | 16.6 | 77.3 | 5.1 | 0.74 | 14,715 | 82.8 | 15,600 |
| 5 | II | 1 | Pa. | Cambria | 1.3 | 17.5 | 70.9 | 10.3 | 1.68 | 13,800 | 81.3 | 15,595 |
| 6 | II | 2 | Pa. | Somerset | 1.5 | 20.8 | 67.5 | 10.2 | 1.68 | 13,720 | 77.5 | 15,485 |
| 7 | II | 2 | Pa. | Indiana | 1.5 | 23.4 | 64.9 | 10.2 | 2.20 | 13,800 | 74.5 | 15,580 |
| 8 | II | 3 | Pa. | Westmoreland | 1.5 | 30.7 | 56.6 | 11.2 | 1.82 | 13,325 | 65.8 | 15,230 |
| 9 | II | 3 | Ky. | Pike | 2.5 | 36.7 | 57.5 | 3.3 | 0.70 | 14,480 | 61.3 | 15,040 |
| 10 | II | 3 | Ohio | Belmont | 3.6 | 40.0 | 47.3 | 9.1 | 4.00 | 12,850 | 55.4 | 14,380 |
| 11 | II | 4 | Ill. | Williamson | 5.8 | 36.2 | 46.3 | 11.7 | 2.70 | 11,910 | 57.3 | 13,710 |
| 12 | II | 4 | Utah | Emery | 5.2 | 38.2 | 50.2 | 6.4 | 0.90 | 12,600 | 57.3 | 13,560 |
| 13 | II | 5 | Ill. | Vermilion | 12.2 | 38.8 | 40.0 | 9.0 | 3.20 | 11,340 | 51.8 | 12,630 |
| 14 | III | 1 | Mont. | Musselshell | 14.1 | 32.2 | 46.7 | 7.0 | 0.43 | 11,140 | 59.0 | 12,075 |
| 15 | III | 2 | Wyo. | Sheridan | 25.0 | 30.5 | 40.8 | 3.7 | 0.30 | 9,345 | 57.5 | 9,745 |
| 16 | III | 3 | Wyo. | Campbell | 31.0 | 31.4 | 32.8 | 4.8 | 0.55 | 8,320 | 51.5 | 8,790 |
| 17 | IV | 1 | N.D. | Mercer | 37.0 | 26.6 | 32.2 | 4.2 | 0.40 | 7,255 | 55.2 | 7,610 |

Notes: For definition of Rank Classification according to ASTM requirements, see Table 3.

Data on Coal (Bed Moisture Basis)

M = equilibrium moisture, %; VM = volatile matter, %;  Rank FC = dry, mineral-matter-free fixed carbon, %;
FC = fixed carbon, %; A = ash, %; S = sulfur, %;  Rank Btu = moist, mineral-matter-free Btu/lb.
Btu = Btu/lb, high heating value.  Calculations by Parr formulas.

**Table 5-2:** Properties of some U.S. coals (Reproduced with permission from *Steam*, 40th ed., published by Babcock & Wilcox, a McDermott Company)

When any coal is combusted in a conventional boiler, most of the sulfur (typically 95%) burns completely with oxygen to form sulfur dioxide. The other 5% of the sulfur exits with the ash and unburned carbon. It does not play a part in this environmental discussion.

$$S + O_2 \rightarrow SO_2$$

| State | Anthracite | Pittsburgh #8 HV Bituminous (Ohio or Pa.) | Illinois #6 HV Bituminous (Illinois) | Upper Freeport MV Bituminous (Pennsylvania) | Spring Creek Subbituminous (Wyoming) | Decker Subbituminous (Montana) | Lignite (North Dakota) | Lignite (S. Hallsville) (Texas) | Lignite (Bryan) (Texas) | Lignite (San Miguel) (Texas) |
|---|---|---|---|---|---|---|---|---|---|---|
| | — | Ohio or Pa. | Illinois | Pennsylvania | Wyoming | Montana | North Dakota | Texas | Texas | Texas |
| **Proximate:** | | | | | | | | | | |
| Moisture | 7.7 | 5.2 | 17.6 | 2.2 | 24.1 | 23.4 | 33.3 | 37.7 | 34.1 | 14.2 |
| Volatile matter, dry | 6.4 | 40.2 | 44.2 | 28.1 | 43.1 | 40.8 | 43.6 | 45.2 | 31.5 | 21.2 |
| Fixed carbon, dry | 83.1 | 50.7 | 45.0 | 58.5 | 51.2 | 54.0 | 45.3 | 44.4 | 18.1 | 10.0 |
| Ash, dry | 10.5 | 9.1 | 10.8 | 13.4 | 5.7 | 5.2 | 11.1 | 10.4 | 50.4 | 68.8 |
| **Heating value, Btu/lb:** | | | | | | | | | | |
| As-received | 11,890 | 12,540 | 10,300 | 12,970 | 9,190 | 9,540 | 7,090 | 7,080 | 3,930 | 2,740 |
| Dry | 12,880 | 13,230 | 12,500 | 13,260 | 12,110 | 12,450 | 10,630 | 11,360 | 5,960 | 3,200 |
| MAF | 14,390 | 14,550 | 14,010 | 15,320 | 12,840 | 13,130 | 11,960 | 12,680 | 12,020 | 10,260 |
| **Ultimate:** | | | | | | | | | | |
| Carbon | 83.7 | 74.0 | 69.0 | 74.9 | 70.3 | 72.0 | 63.3 | 66.3 | 33.8 | 18.4 |
| Hydrogen | 1.9 | 5.1 | 4.9 | 4.7 | 5.0 | 5.0 | 4.5 | 4.9 | 3.3 | 2.3 |
| Nitrogen | 0.9 | 1.6 | 1.0 | 1.27 | 0.96 | 0.95 | 1.0 | 1.0 | 0.4 | 0.29 |
| Sulfur | 0.7 | 2.3 | 4.3 | 0.76 | 0.35 | 0.44 | 1.1 | 1.2 | 1.0 | 1.2 |
| Ash | 10.5 | 9.1 | 10.8 | 13.4 | 5.7 | 5.2 | 11.1 | 10.4 | 50.4 | 68.8 |
| Oxygen | 2.3 | 7.9 | 10.0 | 4.97 | 17.69 | 16.41 | 19.0 | 16.2 | 11.1 | 9.01 |
| **Ash fusion temps, F** | | | | | | | | | | |
| Reducing/Oxidizing (Red Oxid) | — — | 2220 2560 | 1930 2140 | 2750+ 2750+ | 2100 2180 | 2120 2420 | 2030 2160 | 2000 2210 | 2370 2470 | 2730 2750+ |
| ST Sp. | — — | 2440 2640 | 2040 2330 | " " | 2160 2300 | 2250 2470 | 2130 2190 | 2060 2250 | 2580 2670 | 2750+ " |
| ST Hsp. | — — | 2470 2650 | 2080 2400 | " " | 2170 2320 | 2270 2490 | 2170 2220 | 2090 2280 | 2690 2760 | |
| FT 0.0625 in. | — — | 2570 2670 | 2420 2600 | " ' | 2190 2360 | 2310 2510 | 2210 2280 | 2220 2350 | 2900+ 2900+ | " " |
| FT Flat | — — | 2750+ 2750+ | 2490 2700 | " ' | 2370 2700 | 2380 2750+ | 2300 2300 | 2330 2400 | 2900+ 2900+ | " " |
| **Ash analysis:** | | | | | | | | | | |
| $SiO_2$ | 51.0 | 50.58 | 41.68 | 59.60 | 32.61 | 23.77 | 29.80 | 23.32 | 62.4 | 66.85 |
| $Al_2O_3$ | 34.0 | 24.62 | 20.6 | 27.42 | 13.38 | 15.79 | 15.0 | 13.0 | 21.5 | 23.62 |
| $Fe_2O_3$ | 3.5 | 17.16 | 19.0 | 4.67 | 7.53 | 6.41 | 9.0 | 22.0 | 3.0 | 1.18 |
| $TiO_2$ | 2.4 | 1.10 | 0.8 | 1.34 | 1.57 | 1.08 | 0.4 | 0.8 | 0.5 | 1.46 |
| $CaO$ | 0.6 | 1.13 | 8.0 | 0.62 | 15.12 | 21.85 | 19.0 | 22.0 | 3.0 | 1.76 |
| $MgO$ | 0.3 | 0.62 | 0.8 | 0.75 | 4.26 | 3.11 | 5.0 | 5.0 | 1.2 | 0.42 |
| $Na_2O$ | 0.74 | 0.39 | 1.62 | 0.42 | 7.41 | 6.20 | 5.80 | 1.05 | 0.59 | 1.67 |
| $K_2O$ | 2.65 | 1.99 | 1.63 | 2.47 | 0.87 | 0.57 | 0.49 | 0.27 | 0.92 | 1.57 |
| $P_2O_5$ | — | 0.39 | — | 0.42 | 0.44 | 0.99 | — | — | — | — |
| $SO_3$ | 1.38 | 1.11 | 4.41 | 0.99 | 14.56 | 18.85 | 20.85 | 9.08 | 3.50 | 1.32 |

Note: HV = high volatile; MV = medium volatile; ID = initial deformation temp; ST = softening temp; FT = fluid temp; Sp. = spherical; Hsp. = hemispherical.

**Table 5-3:** Properties of U.S. coals including ash analyses (Reproduced with permission from *Steam*, 40th ed., published by Babcock & Wilcox, a McDermott Company)

A small amount of the sulfur dioxide (1% to 4%) further oxidizes to sulfur trioxide ($SO_3$) in the boiler, and the remaining $SO_2$, if left untreated, escapes with the boiler flue gas. Some of the $SO_2$ that enters the atmosphere also oxidizes to $SO_3$. $SO_3$ is the anhydrous form of sulfuric acid, which is a primary precursor of the "acid rain" phenomenon. Sulfur oxides also produce or influence the production of very fine particulates and aerosols.

Figure 5-2 illustrates the strategies adopted by utility managers to meet the $SO_2$ regulations for the CAAA of 1990. Fuel switching was by far the most dominant choice for Phase I compliance, in large part because it represented the least expensive and quickest fix. However, as detailed in Appendix 3-2, fuel switching does not come without problems.

Allowance trading is another common technique utilities use to meet $SO_2$ emissions guidelines. The 1990 CAAA incorporated provisions by which each utility was allowed a certain amount of $SO_2$ emissions (tons) per year. If the utility produced less than its limit of $SO_2$, the company received $SO_2$ "credits," each credit counting as one ton of $SO_2$ per year. The EPA allowed utilities to sell these credits up to the point of its emissions limits. Since the early 1990s, utilities that need to comply with $SO_2$ control limits have purchased $SO_2$ emissions credits from facilities controlling $SO_2$ below mandated guidelines.

| Method | No. of Units | % of Units | % SO$_2$ Reduction from 1985 Baseline | % of Total SO$_2$ Reduction |
|---|---|---|---|---|
| Fuel switching/blending | 136 | 52 | 60 | 59 |
| Additional SO$_2$ allowances | 83 | 32 | 16 | 9[a] |
| Scrubbers | 27 | 10 | 83 | 28 |
| Retirements | 7 | 3 | 100 | 2 |
| Other[b] | 8 | 3 | 86 | 2 |
| Total | 261 | 100 | 345 | 100 |

[a] Includes reduced coal consumption of 2.5 million tons and 16% reduction in sulfur content.
[b] Includes 1 repowered unit, 2 switched to natural gas, and 5 switched to No. 6 fuel oil. Source: *The Effects of Title IV of the Clean Air Act Amendments of 1990 on Electric Utilities: An Update,* Energy Information Administration, March 1997.

**Fig. 5-2:** Phase I SO$_2$ compliance methods (Source: *Clean Coal Technology Demonstration Program: Program Update 2000;* U.S. Department of Energy, Washington, D.C., April 2001)

Allowance trading has been a major, but temporary method by which utilities can meet Phase II compliance requirements. With ever-tightening regulations, many utility managers will have to examine additional control methods. Even fuel switching may not be a complete answer. Flue gas scrubbing is probably the most complete solution, and the EPA projects the number of scrubbers will dramatically increase. Polishing scrubbers may be needed on units that presently emit low levels of SO$_2$ but will not be in compliance with future guidelines.

The most popular scrubbing process has been wet-limestone flue gas desulfurization. A basic design is illustrated in Figure 5-3. The process involves the following steps.

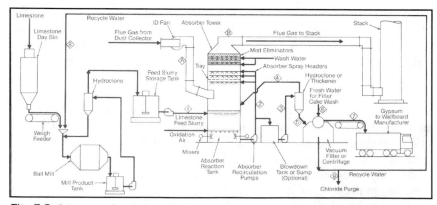

**Fig. 5-3:** Schematic of a wet-limestone flue gas desulfurization system (Reproduced with permission from *Steam,* 40th ed., published by Babcock & Wilcox, a McDermott Company)

Sulfur dioxide is soluble in water. When the flue gas contacts a liquid spray, sulfur dioxide absorbs into the liquid:

$$SO_2 + H_2O \rightleftharpoons SO_2 \cdot H_2O$$

This reaction is reversible. This means that $SO_2$ will evolve from solution unless further reactions take place. However, as $SO_2$ absorbs into water, acidity increases, and the acidic liquid will then react with alkaline materials. Limestone is an inexpensive, natural, alkaline material. The principal alkaline compound in limestone is calcium carbonate ($CaCO_3$). When it is added to the scrubbing solution, the following reaction takes place:

$$SO_2 + \tfrac{1}{2}H_2O + CaCO_3 \rightarrow CaSO_3 \cdot \tfrac{1}{2}H_2O\downarrow + CO_2 \uparrow$$

In Figure 5-3, limestone is ground and mixed with water in a wet ball mill. The resulting slurry is injected into an absorber tower. The absorber recirculation pumps push the slurry out of spray nozzles in the upper portion of the tower, where the spray contacts the rising flue gas. The partially reacted slurry falls to the bottom of the tower and is promptly returned to the spray nozzles. The reaction product, calcium sulfite hemihydrate ($CaSO_3 \cdot 0.5H_2O$) forms a precipitate that may be disposed directly. However, this compound exists as a sludge that retains water and is difficult to transport. Many facilities include an additional oxidation step:

$$2CaSO_3 \cdot \tfrac{1}{2}H_2O + O_2 + 1.5H_2O \rightarrow 2CaSO_4 \cdot 2H_2O\downarrow$$

The reaction produces gypsum ($CaSO_4 \cdot 2H_2O$), the material used in wallboard. This compound is much easier to dewater and can be landfilled. A number of utilities produce gypsum of wallboard quality and sell it to manufacturers. In Figure 5-3, oxidation air is injected into the slurry at the bottom of the absorber tower. A side stream takes reacted material to the blowdown sump from which the slurry is withdrawn for filtering and disposal.

These equations are simplified, as many of the reactions, including the oxidation of calcium sulfite to gypsum, take place in the liquid phase. However, they do accurately represent the overall process.

Of primary importance in wet-limestone scrubbing is adequate removal of $SO_2$ from the flue gas and good utilization of the limestone. Several factors influence these processes, including contact between the flue gas and limestone slurry, reactivity of the limestone reagent, and particle size. First- and second-generation scrubbers were usually equipped with internal packing or trays to enhance gas-liquid contact. This material would often become covered and clogged with scale, requiring periodic cleaning, replacement, or laborious scale-control methods.

However, spray technology has improved since then and open spray towers are now becoming popular; trays are no longer necessary. The spray nozzles produce very fine droplets that contact and mix with virtually all of the incoming flue gas.

As this author can attest from his own experience as a flue gas desulfurization engineer, limestone reactivity is another key factor. Some limestones, and particularly dolomitic limestones that contain significant quantities of magnesium carbonate ($MgCO_3$), may react slowly with the acidic solution. This can require an excess of limestone reagent to achieve the required $SO_2$ removal. Additives have been developed to enhance performance. One of the most popular of these goes by the common name of dibasic acid (DBA). DBA is an organic compound with two acid-producing groups. DBA assists limestone dissolution and consequent $SO_2$ removal. Automatic DBA feed systems are a practical approach for enhancing the $SO_2$-removal performance of existing scrubbers.

Limestone reactivity is greatly influenced by particle size. A typical method of preparing limestone slurry is to grind coarse limestone with water in a ball mill. This produces a suspended solution of fine limestone particles (slurry), which is then pumped to the reaction vessel. Smaller particle size increases the total surface area of the limestone reactant. This enhances the contact with the acidic solution, which not only speeds up $SO_2$ removal but also increases limestone consumption. A well-run scrubber may see $SO_2$ removal of up to 98% with high limestone utilization values. A key is to establish grinding mill conditions so maximum reactivity is achieved without excessive grinding of the limestone.

A scrubber system requires considerable space. The scrubbing vessel itself may be 20 to 30 feet in diameter and perhaps 70 to 100 feet high. Raw limestone storage and handling systems, limestone grinding and gypsum dewatering equipment, and gypsum storage pads take up considerably more room. When the byproduct must be disposed in landfills, land requirements become much more substantial.

A modification to wet-limestone scrubbing is wet-lime scrubbing. Reagent feed requires that the pebbled lime be slaked with water to produce calcium hydroxide [$Ca(OH)_2$], which is then fed to the absorber tower. The major difference is that the reagent is pebbled lime (calcium oxide, CaO). Because lime is much more reactive than limestone, a smaller scrubber vessel will achieve the same $SO_2$ removal. Counterbalancing these advantages is that lime is a reactive material, and fugitive dust emissions represent a safety issue. Slaking systems are often difficult to manage, can be a source of mechanical difficulties, and can release reactive lime to the surroundings. Limestone-based systems have been the more popular choice of wet scrubbers.

Another scrubbing technique is the spray dryer. A generic outline is shown in Figure 5-4. In this method, wet lime (the preferred reagent) is sprayed into the flue gas. The moisture concentration is limited, so the heat from the flue gas evaporates the water and the particles are collected as a dry material, usually in a baghouse. Lime is preferable to limestone due to its greater reactivity. One potential drawback to dry scrubbing is that the $SO_2$ reacts from the outside in. This may coat the particles with a layer of calcium sulfite and sulfate, leaving un-reacted material at the center. A trade-off is the smaller scrubber size and the potential to more easily retrofit these units to existing steam generators. This offers the possibility of using these units as finishing scrubbers for boilers in which sulfur dioxide production is already low, but will not be in compliance with tough future guidelines. Dry scrubbers also offer potential as a mercury control method. This aspect will be outlined later.

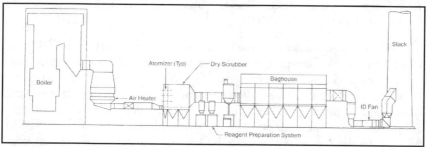

**Fig. 5-4:** Schematic of a dry scrubbing system (Reproduced with permission from *Steam*, 40th ed., published by Babcock & Wilcox, a McDermott Company)

A drawback to dry scrubbing is that the flyash becomes laden with calcium sulfite and sulfate, which may make it unsuitable for use in traditional waste recycling processes such as road fill. Landfilling may become the only alternative for disposal.

# CONTROL OF NITROGEN OXIDES

Another major ingredient of "acid rain" (and a troublesome pollutant for other reasons) is $NO_x$. The term $NO_x$ is all-inclusive, as fossil-fuel combustion produces several oxides of nitrogen, most notably nitrogen oxide ($NO_x$) and nitrogen dioxide ($NO_2$). A typical ratio of $NO_x$ to $NO_2$ is 9 to 1. In plants equipped with a sulfur dioxide scrubber, some $NO_2$ will wash out of the flue gas. However, NO is only slightly soluble, and since it usually makes up about 90% of the $NO_x$, scrubbing is not an effective treatment process.

| Technology | Approzimate discharge concentration (ppmv) at 3% excess oxygen | Approximate discharge concentration (lb/MBtu) at 3% excess oxygen |
|---|---|---|
| Standard Burners | 120 | 0.14 |
| Low NO$_x$ Burners | 45 | 0.06 |
| Current Ultra Low NO$_x$ Burners | 15 | 0.018 |
| Future Ultra Low NO$_x$ Burners | 6 | 0.007 |
| Flue Gas Recirculation | 20 | 0.025 |
| Selective Non-Catalytic Reduction | 27-70 | 0.033-0.085 |
| Sekectuve Catakttuc Reduction | 5-12 | 0.006-0.015 |
| Data aadapted from Reference 3 | | |

**Table 5-4**: NO$_x$ reduction techniques (Source: Bradford, P.E., M and R. Graver, "Just Say NO$_x$." *Environmental Protection*, December 2000.)

When nitrogen oxides enter the atmosphere, several problems arise. NO$_x$ participates in a series of complex photochemical reactions with volatile organic compounds that produce ground-level ozone (O$_3$). While the ozone layer at high levels in the earth's atmosphere protects us from harmful radiation, ground-level ozone causes respiratory problems, especially in the young, the elderly, smokers, asthmatics, and people with other lung problems. Upon release from the boiler, much of the nitrogen oxide converts to NO$_2$, which like its SO$_2$ counterpart, combines with water to form an acid—in this case, nitric acid (HNO$_3$). Nitrogen oxides also contribute to the production of very fine particulates and aerosols.

NO$_x$ is the only emission from natural gas-fired plants and combined-cycle units that is currently of major concern to environmental regulators. Nitrogen oxides produced during coal combustion are generally grouped into two categories—thermal NO$_x$ and fuel NO$_x$.

Thermal NO$_x$ is generated by the high heat of combustion, and results from the reaction of atmospheric nitrogen (N$_2$) and oxygen (O$_2$). Even in pulverized coal units, thermal NO$_x$ may only account for a quarter of all NO$_x$ emissions, as thermal NO$_x$ formation does not really develop until temperatures reach 2,800°F (1538°C). Such high energy is needed to break the double-bonded nitrogen molecules. A popular technique to reduce thermal NO$_x$ is flue gas recirculation, in which a portion of the flue gas is recycled to the boiler. This lowers furnace temperatures. Increasing the size of the combustion zone for a given thermal input also reduces temperature.

Fuel $NO_x$ is a different story. Most nitrogen found in coal is organically bound to carbon atoms, and these bonds are much easier to break. As the coal combusts, individual nitrogen atoms are released. Nitrogen atoms (as opposed to molecules) are very reactive and quickly attach to oxygen.

Table 5-4 outlines the most popular $NO_x$ reduction methods and the practical limits at which they perform. Techniques and equipment include low $NO_x$ burners, overfire air, selective catalytic reduction, and selective non-catalytic reduction. The next sections examine the fundamentals of these processes.

# Low-$NO_x$ burners and overfire air

Control of nitrogen oxide formation and discharge generally falls into two categories—concurrent combustion control and post combustion control.

Low-$NO_x$ burners (LNB) and overfire air (OFA) belong to the first category. The names of these two techniques do not hint at the fact that processes are influenced by chemistry, but in actuality, they are. The detailed chemistry is complex, but the overall process is enlightening. When coal or any other fossil fuel is burned with an excess of oxygen, combustion of the carbon content proceeds to completion as follows:

$$C + O_2 \rightarrow CO_2 \uparrow$$

Energy-wise, this reaction is very favorable and gives off much heat in the process. When sub-stoichiometric amounts of oxygen are fed to the process, a fraction of the coal remains unburned and some only partially oxides to carbon monoxide:

$$C + \frac{1}{2}O_2 \rightarrow CO \uparrow$$

The unburned coal and carbon monoxide both seek oxygen atoms to complete the reaction to carbon dioxide. When insufficient oxygen is available, the molecules will take oxygen from nitrogen oxides. This is the basis behind LNB and OFA—the fuel is initially combusted in an oxygen-lean environment to allow the formation of some reduced carbon species. These reduced carbon compounds strip oxygen from $NO_x$ and allow nitrogen atoms to combine into their most thermodynamically favored structure, $N_2$.

The chemistry becomes complex because of the huge number of molecular interactions that take place, even during the short time that the molecules are in the furnace. A nitrogen oxide molecule may give up its oxygen atom(s) to carbon only to combine with other oxygen. This process may happen repeatedly before the nitrogen atom meets another nitrogen atom to form $N_2$. Many chemical species are produced during combustion, and the interactions that eventually reduce $NO_x$ levels are quite complex. The remaining air for combustion is added after these reactions have taken place to convert the remaining unburned carbon and carbon monoxide to $CO_2$.

A low-$NO_x$ burner schematic is shown in Figure 5-5. Fuel, lean in air, flows out of the center of the burner. It is in this zone that reducing reactions take place. The remaining air flows through secondary zones at the perimeter of the burner. This completes the combustion process but not until $NO_x$ production has been greatly reduced. Low-$NO_x$ burners of this type serve as retrofit equipment for wall-fired and opposed-fired boilers. Ultra $NO_x$ burners represent the second and third generation of low $NO_x$ combustors. Table 5-4 illustrates how current and future models produce or are expected to produce low emissions.

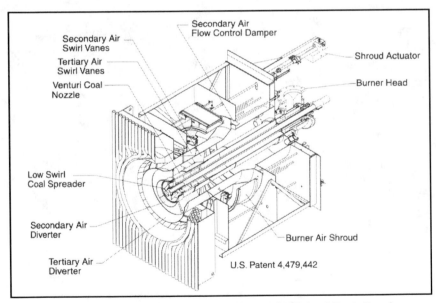

**Fig. 5-5:** Diagram of a low-$NO_x$ burner (Illustration provided by Babcock Borsig Power)

A supplement to LNB is OFA. In this technique, some of the air needed for complete combustion is introduced above the burners. This provides more time for reduced carbon species to interact with nitrogen oxides and generate elemental nitrogen. OFA air can be used on many types of boilers, including stokers, to lower $NO_x$.

The LNB/OFA air technique is proving to be very important on the most popular type of coal units, pulverized-coal, tangentially-fired boilers. Figure 5-6 shows a generic arrangement of the configuration on a DOE project. As with previous generations of these boilers, the burners are placed in the corners of the boiler to generate a swirling fireball. Modern designs stage the fuel and air feeds to establish reducing conditions during the initial combustion process (Fig. 5-7). The LNB/OFA method has been able to reduce $NO_x$ emissions below 0.20 lb/$10^6$ Btu (0.09 kg/$10^6$ kJ) in tangentially-fired boilers. Similar results have been reported with LNB/OFA on wall-fired boilers.

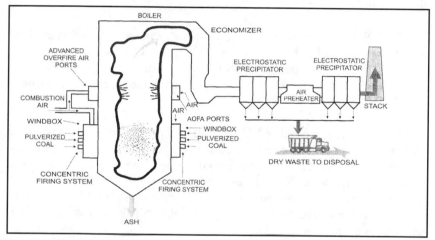

**Fig. 5-6:** Schematic of an overfire air design for a tangentially-fired boiler (Source: *Clean Coal Technology Demonstration Program: Program Update 2000*; U.S. Department of Energy, Washington, D.C., April 2001)

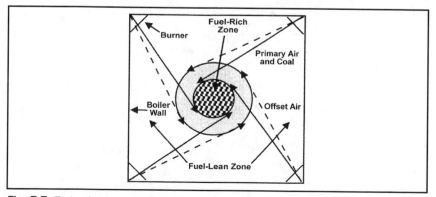

**Fig. 5-7:** Fuel and air zones in a boiler with OFA configuration shown in Figure 5-6 (Source: *Clean Coal Technology Demonstration Program: Program Update 2000*; U.S. Department of Energy, Washington, D.C., April 2001)

A problem with overfire air is that the technique creates a zone of reducing conditions in the boiler. During conventional combustion, excess air is injected with the fuel to ensure that virtually all of the carbon in the coal burns to completion. The excess air establishes an oxidizing atmosphere in the furnace, in which the boiler tubes develop an oxide coating. Where OFA is employed, the combustion products between the main burners and the OFA feed points contain reducing compounds including sulfides. These may react with the tube walls to form iron sulfides that are not as protective as the counterpart oxides. Corrosion and spalling of tube material have been known to occur in boilers retrofitted with overfire air systems.

Low-$NO_x$ burners and OFA—either alone or in combination—raise concerns about increased carbon monoxide production. CO is, of course, one of the primary pollutants regulated in the NAAQS. Well-designed systems minimize CO increases, although as will be shown, post combustion methods are available to control carbon monoxide discharge.

Gas re-burning is another $NO_x$-control method that has been adapted as a retrofit on existing boilers. Here, the coal-firing rate is set below original design conditions, and natural gas is injected into the boiler above the main burners. The general diagram from a DOE gas re-burning project is illustrated in Figure 5-8. The natural gas makes up for the heat loss due to reduced coal feed, but it also supplies carbon atoms for reaction with $NO_x$. The DOE has also sponsored programs with coal as the re-burning fuel.

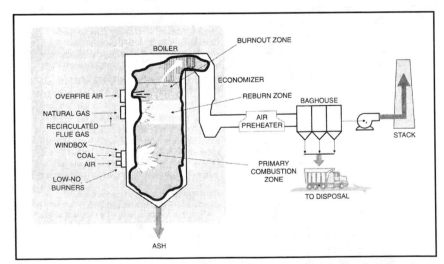

**Fig. 5-8:** Illustration of gas reburning with OFA (Source: *Clean Coal Technology Demonstration Program: Program Update 2000*, U.S. Department of Energy, Washington, D.C., April 2001)

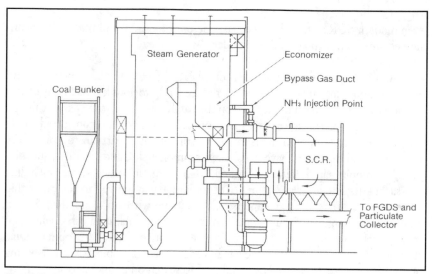

**Fig. 5-9:** Schematic of a steam generator with SCR system (Reproduced with permission from *Combustion: Fossil Power,* published by Alstom)

The combination of low-NO$_x$ burners and overfire air has proven capable of lowering NO$_x$ emissions to levels close to 0.15 lb/MBtu (0.064 kg/$10^6$ kJ). A popular alternative for this *in situ* NO$_x$-control technique is post-combustion control with selective catalytic reduction (SCR) (Fig. 5-9). A cousin to this process is selective non-catalytic reduction (SNCR). Both involve reduction of NO$_x$ by reaction with ammonia, but in the SCR process, the ammonia and flue gas intermingle upon a catalyst bed. Typical SCR reactions are illustrated in the following two equations:

$$4NO + 4NH_3 + O_2 \rightarrow 4N_2\uparrow + 6H_2O$$

$$2NO_2 + 4NH_3 + O_2 \rightarrow 3N_2\uparrow + 6H_2O$$

The key is addition of ammonia to the flue gas, which reacts with NO$_x$ to generate elemental nitrogen. In selective catalytic reduction, the reaction takes place in a fixed catalyst bed mounted in the flue gas stream. A variety of materials are possible for SCR catalysts. Most common are titanium dioxide, vanadium pentoxide, precious metals, and zeolites (aluminosilicates). The ideal operating range of the transition metal catalysts (titanium, vanadium) is generally 450°F to 850°F, while the zeolites operate at a higher temperature range (perhaps 850°F to 1,050°F). The most common structural configuration is "a block-type catalyst manufactured in parallel plate or honeycomb configurations..." These have replaced earlier packed beds, whose performance was subject to fouling. However, even modern SCR systems increase gas-side pressure drop, and this must be considered when retrofitting SCR to a boiler.

SCR introduces several potential complicating factors to plant operations. An excess of ammonia must be added to reduce $NO_x$ to low levels. Some of this ammonia will be oxidized to nitrogen on the catalyst bed, but some passes through the bed unreacted. This is known as ammonia slip. Ammonia is considered a hazardous air pollutant, and it is a precursor of fine particulate formation. Environmental regulators have considered ammonia slip, and "in many jurisdictions, ammonia slip is already limited to 5 ppm; in other states, including Massachusetts, the limit is 2 ppm." Technology has improved to the point that some manufacturers now guarantee SCR systems with no more than 2 ppm ammonia discharge. Such limits are also important to maintain ash quality, and "the general rule from European experience is not to exceed 2-3 ppm residual ammonia concentration in the flue gas in order to avoid problems with disposing or marketing flyash or scrubber byproduct." Research on ammonia oxidation catalysts is underway. If they prove out, they could virtually eliminate ammonia discharges.

SCR catalysts gradually degrade over the life of the material, so periodic replacement of catalysts is necessary. The transition metal catalysts can be poisoned by sulfur and chlorine, so at plants burning high-sulfur coal, zeolite may be the best alternative. Arsenic and calcium oxide also act as catalyst poisons, and these should be taken into account during design or if a utility switches fuels. Disposal of spent transition metal catalysts may be expensive due to the need to control discharge metals to the environment.

Reaction of ammonia with sulfur trioxide in the flue gas produces ammonium sulfate [$(NH_4)_2SO_4$] and ammonium bisulfate ($NH_4HSO_4$), both of which will foul and corrode downstream equipment including air heaters. This is especially true for bisulfate. "The use of enameled air-heater surfaces is standard practice in Europe, and, along with low flue-gas $SO_3$ content, is a key reason why such deposits are not a problem for European applications." The catalyst also increases the conversion rate of $SO_2$ to $SO_3$, and so "one conclusion was that when operating with medium- to high-sulfur coals on high-efficiency plants, $SO_2$ to $SO_3$ conversion and subsequent acid-enhanced deposition could produce a greater fouling risk than ammonia slip and [ammonium bisulfate] formation."

Another problem with ammonia derives from its storage on plant grounds. "The EPA classifies anhydrous ammonia and aqueous ammonia as low as 20% concentrations as regulated toxic substances." Some utilities store 19% concentrations to avoid this issue.

A technique that is becoming popular to generate ammonia for SCR systems is hydrolysis of solid urea. Urea (Fig. 5-10) is a common agricultural chemical that can be stored and handled safely, but can be hydrolyzed on-site to produce ammonia and carbon dioxide. These systems allow ammonia to be generated on demand without storing bulk concentrations of the hazardous material.

$$O$$
$$\|$$
$$H_2N - C - NH_2$$

**Fig. 5-10:** The chemical structure of urea

The reaction of ammonia with $NO_x$ will also proceed without the aid of a catalyst when the temperature is higher, typically within a range of 1,650°F to 1,800°F (899°C to 982°C). This is the basis of SNCR, where no catalyst is used and the ammonia is injected into the upper reaches of the boiler (Fig. 5-11). An obvious advantage to this technique is the lack of a catalyst bed, which offers potential for SNCR to serve as a retrofit $NO_x$-reduction method.

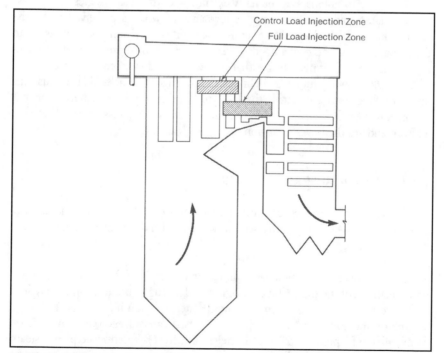

**Fig. 5-11:** Typical locations for SNCR in a coal-fired boiler (Reproduced with permission from *Steam*, 40th ed., published by Babcock & Wilcox, a McDermott Company)

As with SCR, ammonia storage and handling are issues, but again urea serves as a substitute. In this process, the urea may be injected directly into the furnace. A typical reaction is as follows:

$$CO(NH_2)_2 + 2NO + \tfrac{1}{2}O_2 \rightarrow 2N_2\uparrow + CO_2\uparrow + 2H_2O$$

A principal concern with SNCR is that the process may produce ammonia-related fine particulates. This could run afoul of PM2.5 regulations.

$NO_x$ conversion rates in SCR systems may reach or exceed 90% and, in some cases, SCR can serve as stand-alone $NO_x$ control devices. (SNCR alone may only be half as efficient.) In other cases, a combination of techniques, such as low-$NO_x$ burners and SCR, may be best. These decisions must be made at the individual plant level, as they depend upon the type of combustion process and the nature of the fuel. For the rebuild of Kansas City Power and Light Company's Hawthorn No. 5 unit, the plant designers selected LNB, OFA, and SCR. This combination is expected to keep $NO_x$ emissions below 0.10 lb/MBtu.

One item to note is that the EPA examines control technologies closely and often updates guidelines based on performance of what is perceived to be the "best" solution. If the EPA sets an emissions limit of a certain pollutant as "x" in the guidelines, and the regulators discover another practical process that produces lower emissions, they may decide that is the "best available control technology" (BACT) and modify future guidelines to meet this criteria. BACT evaluations take economics into account. If the EPA feels air quality is extremely poor in a region, a plant may be subjected to the lowest achievable emission rate (LAER), which mandates the lowest emissions possible regardless of cost.

# $NO_x$ control for gas turbines

This book primarily covers steam generation topics, but because gas turbines power combined-cycle units, this section will address $NO_x$ control in combustion turbines.

$NO_x$ produced by natural gas combustion in simple-cycle or combined-cycle power plants is of the thermal type. Logic therefore suggests a technique to reduce $NO_x$ is to lower combustion temperatures. Water injection into the gas turbine is a common method, and "water injection by itself can reduce $NO_x$ emissions on many units to 25 ppm at 15% $O_2$ [in the stack]." In combined-cycle plants, steam injection is also possible. Besides lowering combustion temperatures, water or steam injection also increases combustion turbine capacity, as the output is influenced by the mass flow rate through the turbine.

Regulatory authorities realize, however, that it is possible to reduce $NO_x$ levels much lower than is possible with water/steam injection, and so "among the many forces driving [$NO_x$ limits] downward is the regional ozone problem." An important technique is staged combustion—similar in theory to the coal-fired examples mentioned above—in which the fuel is first combusted in an air-lean environment to reduce $NO_x$ formation, followed by the remaining air feed. Single-digit $NO_x$ levels (in parts-per-million) are possible with these combustors. Further, gas turbine service firms have developed combustor replacement components that may easily be retrofitted to existing units to lower $NO_x$ levels to single digit values.

Another emerging technique is that of catalytic combustion. In this process, the fuel and air are burned flamelessly in a catalyst bed. The maximum combustion temperature is around 1,800°F (982°C). This lowers thermal $NO_x$ formation. At the time of this publication, the catalytic combustion technique is still in the testing stage, but $NO_x$ emissions of less than 3 ppm (with carbon monoxide emissions of less than 10 ppm) have been demonstrated.

SCR is a proven control method, and has been adapted to both simple-cycle and combined-cycle units. Simple-cycle installation is relatively easy, as the ammonia injection system and catalyst bed reside downstream of the turbine with no heat exchangers further downstream to become fouled. In a combined-cycle unit, the ammonia injection system and catalyst bed usually reside between heat exchange circuits, where the temperature is within the window for necessary chemical reactions. While this can potentially subject downstream equipment to catalyst-bed carryover, ammonium sulfate and bisulfate formation is not of concern, unlike coal-fired boilers. Problems may arise, however, when dual-fuel units are switched from natural gas to fuel oil. Particulate and sulfur loading of the catalyst bed then become an issue, as does the possibility of ammonium bisulfate fouling of the low-pressure economizer.

Multiple technologies are often used to control $NO_x$. For example, a technique possible with SCR systems is to add a separate catalyst bed to convert carbon monoxide into $CO_2$. Equipment in a modern coal-fired boiler might include low-$NO_x$ burners, followed by SCR, followed by catalyst-reduction of carbon monoxide. The economics might very well justify multiple arrangements as well as meeting potential EPA/BACT criteria. In this scenario, the low-$NO_x$ burners would reduce $NO_x$ emissions by 50%, significantly lowering the load on the SCR. Additional carbon monoxide produced by the burners would be eliminated in the CO-catalyst bed. Overfire air could always be included in this scenario to further reduce $NO_x$ load on the SCR.

## Particulate control

The type of boiler greatly influences particulate formation. Older-style cyclone boilers were designed to send approximately 80% of the ash out through the bottom of the boiler in a molten state. In pulverized-coal units (where typically 80% of the ash exits with the flue gas), particulate discharge may be quite high. Existing requirements at all plants require capture of most (99%) of these particulates. Two major processes—electrostatic precipitation (ESP) and fabric filter collection, with supplemental variations—are popular for particulate control. However, impending regulations may require additional or more advanced controls.

A simple view of the ESP process is shown in Figure 5-12. This is a plate and weighted-wire type unit. An electric potential exists between the plates and wires, where a negative potential is applied to the wires and a positive potential to the plates. The potential difference typically ranges between 45 kilovolts (kV) and 80 kV. The negative electrodes resemble barbed wire with points protruding along the length. As the flue gas passes through the precipitator, the particles accumulate electrical charge from the wire points. The charged particles then migrate to the plates, which are periodically shaken (rapped) by mechanical vibrators. The ash falls to the bottom of the precipitator and is collected in hoppers for discharge via the ash disposal system.

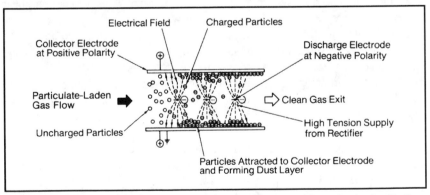

**Fig. 5-12:** Principles of ESP operation (Reproduced with permission from *Combustion: Fossil Power*, published by Alstom)

The flue gas duct opens out into the much larger ESP. This design slows the linear velocity of the gas so the particles have more time to develop a charge. A typical entering velocity might be 60 feet per second (18.3 meters per second), while flow through the precipitator might only be 4 to 5 feet per second (1.3 to 1.7 mps). ESPs consist of a series of cells, where ash removal efficiency is around 75% per each cell. When combined in series, overall efficiencies of 99% are possible.

The distance between collecting plates in an ESP may range from 9 to 16 inches, with the negative electrode evenly centered between collecting plates. The weighted-wire design has fallen out of use in favor of more durable arrangements, where the charging electrode is more solid and break resistant. Figure 5-13 shows one of these newer designs—the rigid-frame type.

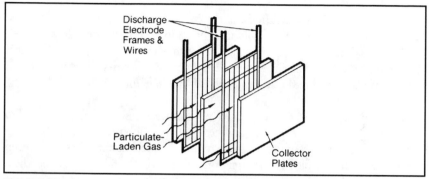

**Fig. 5-13:** Outline of a rigid frame ESP (Reproduced with permission from *Combustion: Fossil Power*, published by Alstom)

A number of factors influence precipitator performance. Most notably, particulate removal efficiency is dependent upon the nature of the flyash and, in particular, its resistivity. Resistivity is a measure of the ability of ash particles to accept a charge. Figure 5-14 illustrates ash resistivities for various coals as a function of temperature.

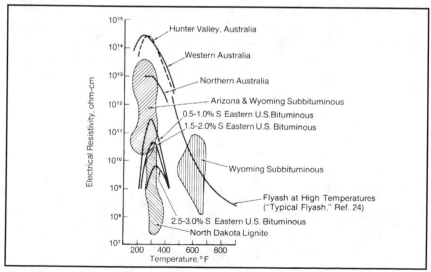

**Fig. 5-14:** Ash resistivities of various coals as a function of temperature (Reproduced with permission from *Combustion: Fossil Power*, published by Alstom)

*Experience has shown that coal ashes with resistivities above 5 x 10$^{11}$ ohm-cm are difficult to collect. Low resistivity (less than 10$^9$ ohm-cm) ash tends to suffer from excessive reentrainment, but such low resistivities are rarely encountered. The area between 10$^9$ and 5 x 10$^{11}$ ohm-cm is usually quite normal for satisfactory and predictable precipitator design.*

As Figure 5-14 illustrates, most of the coals used for power generation in the U.S. have ash resistivities that fall within this range. Temperature does have an effect on resistivity, which can influence where the precipitator is located. In most cases, the ESP follows the air heater and is known as a "cold-side" precipitator. If the ESP is placed before the air heater, it is a "hot-side" precipitator.

ESPs are susceptible to performance problems. Excessive rapping may lead to re-entrainment of ash particles in the flue gas. Poor combustion in the boiler can cause excessive buildups of unburned carbon in the precipitator. These have been known to start fires. Cold-side precipitators are at the back end of the boiler where flue gas temperatures are lowest. If the temperature drops too low in the precipitator, acidic condensation and corrosion are the result. A change in ash quality or improper vibration frequencies will lead to a build-up of ash on the collecting wires, which reduces precipitator efficiency.

Changes in flue gas chemistry can affect precipitator performance—sodium in flue gas enhances performance, a change in fuel to a lower-sodium coal may upset operation, small quantities of sulfur trioxide ($SO_3$) produced during combustion also enhance precipitator performance. A switch to lower-sulfur coal can potentially affect ESP performance, while excess ammonia carryover from a SCR or SNCR will react with sulfur trioxide and remove it from the flue gas.

The most important factors affecting ESP performance are summarized below.

### Fundamental factors affecting performance

- Ash resistivity
- Particle size
- Inadequate rapping system
- Insufficient transformer/rectifier sets (the devices that convert AC electricity to DC)
- Undersized ESP
- Unstable transformer/rectifier controls

### Mechanical factors affecting performance

- Poor electrode alignment
- Vibrating or swinging electrode
- Distorted collecting plates
- Excessive ash deposits on plates or wires
- Full hoppers
- Casing leaks
- Plugged gas distribution devices

### Operational factors affecting performance

- Improper transformer/rectifier adjustment
- Excessive gas flow
- Process upsets
- Improper adjustment of rapper intensity or frequency

A twist to power industry electrostatic particulate control is the wet ESP (WESP).

*For more than 50 years…wet ESPs have been standard technology in sulfuric acid plants to abate $H_2SO_4$ mist, a sub-micron droplet…Though a typical industrial power boiler's air volume is only 50,000–150,000 cfm [cubic-feet-per-minute, or 1416–4248 cubic-meters-per second], and a typical utility [steam generator] has an airflow of approximately 500,000 to 1,500,000 cfm [14,158–42,475 cmm], the chemistry of the pollutants and the particles being emitted are similar.*

The ESP and WESP processes differ in that WESP temperatures are kept low to condense water vapor from the flue gas. The liquid continually washes the collecting plates, eliminating the need for mechanical rapping. Also, wet electrostatic precipitators may be wire-and-plate or tubular in design.

The WESP process offers several advantages over dry ESP. First, collected particles do not tend to re-entrain as they do when rapped in a standard ESP. The re-entrainment phenomenon affects the size of the ESP, as each downstream cell has to handle more particulates than if all were discharged to the hoppers during rapping. Also, in dry units, the build-up of particulates on the collector plates affects the potential difference between wire-and-plate and the collection efficiency. Thus, fine control of rapping frequency is necessary to prevent excessive re-entrainment, but also to prevent excessive ash buildups and loss of collection efficiency. Continual washing in a WESP eliminates that difficulty.

One wet ESP application that shows promise is as a particulate control device after wet flue gas desulfurization systems. Even though scrubbers do a very good job of removing $SO_2$ from flue gas, they tend to emit particulates. In a conventional steam-generating system with a dry-ESP and FGD system, the scrubber always comes after the ESP. While scrubbers are equipped with stationary mist eliminators at the top of the scrubbing vessel, some particulates still escape. The WESP promises better particulate control methods for scrubber emissions. Researchers are conducting intensive field testing of this process as well as WESP use for direct particulate removal from flue gas.

Another common, and increasingly popular, particulate removal technique is fabric filtration—commonly known as baghouses. Operation is relatively straightforward. The flue gas passes through a fine mesh fabric filter and particulates collect on one side of the filter. Periodic vibrations dislodge the ash, which falls into hoppers for collection. Figure 5-15 shows the generic outline of a popular type of fabric filtration, known as the pulse-jet fabric filter. The bags are mounted on wire frames within the vessel. Flue gas enters from the side, is deflected by and flows around a baffle plate, then passes through the bags. Flyash remains on the exterior of the bags as the flue gas exits out the top of the vessel. Periodically, a brief impulse of air is blown through the bags to dislodge the ash, which falls to hoppers below.

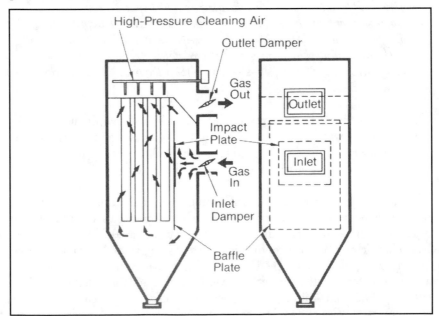

**Fig. 5-15:** Schematic of a pulse-jet fabric filter (Reproduced with permission from *Combustion: Fossil Power*, published by Alstom)

Another type of fabric filter design proven popular is the reverse air type, in which the flow of the flue gas is periodically reversed for short periods via damper gates to dislodge accumulated ash.

Baghouses are typically set up with several compartments. Each contains rows of filters. Redundancy is built in, whereby individual compartments are automatically isolated from the flue gas steam followed by pulse cleaning. This prevents the ash from being re-entrained in the flue gas. Factors that baghouse design engineers must take into account include cloth-to-air ratio, filter diameter and length, and flue gas temperature. Progressive improvements in cloth design now allow baghouses to serve at temperatures of 500°F.

Table 5-5 shows the chemical and temperature limits of a number of bag materials. Even though some materials can withstand high temperatures, they are still susceptible to fires if unburned carbon accumulates in the cloth. Baghouse fires can be very dangerous, because opening of access doors to combat the fire introduces more oxygen. When plain water is sprayed on the fire, burning coal particles tend to float. Utility managers have begun to look at foam or $CO_2$ suppression systems to combat potential baghouse fires.

| Fiber | Fiber Property | | Chemical Resistance | | Recommended Operating Temp., °F | |
|---|---|---|---|---|---|---|
| | Tensile Strength | Abrasion Resistance | Acids | Bases | Continous | Short-time |
| Cotton | Good | Average | Poor | Excellent | 180 | 225 |
| Polyethylene | Excellent | Good | Excellent | Excellent | 190 | 190 |
| Glass | Excellent | Poor | Good | Poor | 500 | 550 |
| Nylon | Excellent | Excellent | Poor | Excellent | 200 | 250 |
| Dacron* | Excellent | Excellent | Good | Fair | 275 | 325 |
| Acrylic* | Average | Average | Very Good | Fair | 260 | 280 |
| Nomex* | Very Good | Very Good | Fair | Very Good | 400 | 425 |
| Teflon* | Average | Below Average | Excellent | Excellent | 450 | 500 |

*Trademark E.I. DuPont

**Table 5-5:** Properties of various fabric filter materials (Reproduced with permission from *Combustion: Fossil Power*, published by Alstom)

Because the baghouse is a filtering medium instead of an electrical charge process like an ESP, removal efficiency is similar across the spectrum of ash properties. This is the factor that has increased baghouse popularity, and will make them even more popular for future control requirements.

# FUTURE ISSUES

As the introduction to this chapter indicated, fine particulate and mercury control are looming issues. Fine particulate control may influence $SO_2$ and $NO_x$ guidelines, as these are known precursors of aerosols and small particles. However, baghouse technology may solve the problem on new units.

A promising technology for existing units is the COHPAC process, in which a fabric filter is retrofitted into the shell of the last field or two of an ESP. The ESP continues to remove the bulk of particulates, while the baghouse acts as a polishing unit to remove carryover.

Baghouse technology may be the answer to mercury control as well. Mercury is a minor element in coal; during combustion, some of the mercury forms an elemental vapor, while some oxidizes and combines with anions like chlorides. For reasons not completely understood, the coal rank (or source) influences the ratio of elemental to oxidized mercury. The ratio of oxidized mercury to the elemental form appears to be much higher in units burning bituminous coal. Oxidized mercury will wash out in a flue gas desulfurization system, but elemental mercury passes on through. Researchers are looking at several methods for complete mercury removal. One is the addition of sodium sulfide ($Na_2S$) to flue gas scrubbing liquors. Mercury reacts very quickly with sulfur to form mercuric sulfide—a very insoluble compound. The mercuric sulfide will leave with the filtered scrubber solids. Dry scrubbing has also been shown to remove some mercury, although the results vary, depending upon the coal rank and other factors.

Many utilities are not equipped with scrubbers. A method under heavy investigation is the combination of baghouse particulate removal with upstream injection of activated carbon. Elemental mercury absorbs onto the activated carbon particles and is discharged with the remaining flyash.

*Use of activated carbon has been shown to remove more than 90 percent of the mercury in a flue gas stream when used in conjunction with a fabric filter. Additionally, it has been found that by impregnating the activated carbon with sulfur, mercury removal capability can be increased from 93 to 96 percent.*

This technique introduces a couple of potential problems. One is that the activated carbon feed increases the amount of combustible material in the baghouse. Second, the mercury becomes entrained within the flyash, which utility managers may have arranged to sell or deposit in a non-hazardous landfill. The implications of mercury in flyash are still pending.

Also under development are multi-pollutant control strategies. An interesting example is the Powerspan system that combines dry and wet ESP with an intermediate high-energy process oxidizing $NO_x$, $SO_2$, and mercury, so they are more easily removed in the WESP.

Brian Schimmoller, editor of PennWell's *Power Engineering* magazine, says, "Nothing strikes more fear into a power plant operator's heart than the threat of…regulations for greenhouse gas emissions, particularly carbon dioxide ($CO_2$)." This is a key issue facing utility owners in the future. Carbon dioxide control is an extremely politically charged issue that pits those concerned about global warming against industry representatives and many economists. While debate continues over the extent and impact of man-made $CO_2$ on global climate, many people, including a number of scientists, are convinced that the issue is very serious. The list of concerned individuals includes some U.S. government leaders, who have proposed caps on future $CO_2$ emissions. At the time of this publication, no limits have been promulgated in the U.S., but legislation is under consideration in Congress to control $NO_x$, $SO_2$, $CO_2$, and mercury produced by coal-fired power plants. The problem stems from the fact that carbon dioxide production from combustion dwarfs other emissions, with no easy or inexpensive method to scrub $CO_2$ from flue gas. Techniques under consideration for $CO_2$ sequestration include deep well and deep-sea injection.

The EPA has also developed a list of 189 chemical substances that constitute a list of hazardous air pollutants (HAP). Many of these are organic compounds, and are of greater concern to such industries as refineries, petrochemical plants, and others of that nature. However, the HAP list includes several heavy metals found in coal. These include lead, arsenic, cadmium, and antimony. The literature indicates that these are easier to control than mercury, but utilities may have to worry about additional environmental controls in the future.

# Conclusion

Air pollution control is a dynamic field governed not only by science, but also by politics. Utility managers will have to keep track of ever-changing conditions and stay abreast of the latest regulations and technologies.

# Appendix 5-1

## Air Quality Update: Staying Abreast of Changing Regulations

### Marc Karell, P.E.
### *Malcolm Pirnie Inc.*

Air pollution has become a major lifestyle and political issue for the American public. The Public demands both an ample supply of available electricity and a clean environment. To achieve this, the federal government over the past three decades has passed increasingly stringent legislation requiring, in many cases, state-of-the-art air pollution controls on new projects. The cost of obtaining permits to construct new plants or modify existing plants has risen substantially over the years. Staying abreast of current air quality rules, therefore, can save a plant owner or developer much money and time

## Ambient air quality standards

In the 1960s, the federal government recognized that pollutants found naturally in the ambient air could become a threat to public health at elevated concentrations. The U.S. Environmental Protection Agency (EPA) developed maximum allowable ambient concentrations for these "criteria" pollutants. Table A5-1 summarizes the current National Ambient Air Quality Standards (NAAQS) for criteria pollutants. Note that a federal appeals court recently issued a stay on the PM-2.5 ambient air quality standard, remanding it to EPA. The court left the revised one-hour ozone standard in place, but as non-enforceable.

A proposed new cogeneration or utility facility or one undergoing a major modification causing a significant increase in emissions must comply with a number of air quality regulations concerning the achievement or maintenance of attainment with the NAAQS. The goal of the federal rule called Prevention of Significant Deterioration (PSD) is to maintain areas currently in attainment with an ambient air quality standard. A facility must obtain a PSD permit if it is "major"

| Pollutant | Averaging Period | Primary Standard | Secondary Standard |
|---|---|---|---|
| Sulfur dioxide | 12 months | 80 ug/m³ | — |
| | 24 hours | 365 ug/m³ | — |
| | 3 hours | — | 1,300 µg/m³ |
| Carbon monoxide | 8 hours | 10 mg/m³ | 10 mg/m³ |
| | 1 hour | 40 mg/m³ | 40 mg/m³ |
| Ozone | 8 hours (3 year average of 4th highest daily max. average) | 0.08 ppm | 0.08 ppm |
| | 1 hour (1 exceedance allowed annually) | 0.12 ppm | 0.12 ppm |
| Nitrogen oxides | 12 months | 100 µg/m³ | 100 µg/m³ |
| Lead | 3 months | 1.5 µg/m³ | — |
| Total suspended particulates (TSP) | 24 hours | 260 µg/m³ | 150 µg/m³ |
| PM-10 | 12 months | 50 µg/m³ | 50 µg/m³ |
| | 24 hours | 150 µg/m³ | 150 µg/m³ |
| PM-2.5 | 12 months | 15 µg/m³ | 15 µg/m³ |
| | 24 hours | 65 µg/m³ | 65 µg/m³ |

**Table A5-1:** National ambient air quality standards

and plans to increase emissions by more than a certain threshold. For each affected pollutant, the applicant must install best available control technology (BACT), which is defined as the most stringent control technology taking into consideration economic, energy, and environmental factors. The PSD application must contain a BACT analysis, analyzing the most stringent applicable control technology. Table A5-2 lists common available air pollution control technologies for criteria pollutants emitted from combustion equipment. An actual BACT analysis should consider compatibility of control equipment with the specific combustion equipment utilized.

An important issue to consider when designing control equipment to meet BACT or any other air quality requirement is that the desired technology may have an adverse effect on other pollutants. For example, technologies to reduce volatile organic compounds (VOC) and CO by oxidizing these pollutants to $CO_2$ will cause nitrogen compounds to form more $NO_x$, potentially exacerbating a problem with that pollutant. Careful planning and communication with regulatory agencies are the keys to not being caught in a position of good faith undercontrolling of a pollutant by implementing technology to reduce another, which can hold up permit issuance. A facility in Maryland recently had its construction permit revoked—after construction had begun—because of this issue.

In addition, the PSD application must contain an impact study using dispersion modeling to show that the net increase in emissions proposed, together with existing facility emissions and existing background concentration of the pollutant, will not cause its NAAQS to be exceeded or exceed a fraction of the remaining impacts.

For example, a power plant proposed in an attainment area that has the potential to emit a high quantity of $NO_x$, $SO_2$, PM-10, or CO must first demonstrate that the ground-level impact of each pollutant, together with the impacts from other nearby combustion sources, will not exceed each NAAQS. But this is not sufficient. PSD discourages a location from reaching the NAAQS quickly. Therefore, regulatory agencies do not want to see a rise in the total impact of a pollutant in an area by more than a fraction or "PSD increment" of the NAAQS over a period of time. To achieve this, regulatory agencies often restrict each applicant for a new facility or major plant modification to a given fraction of the PSD increment.

This has many implications for new power plants, particularly in areas where deregulation is occurring and where building new plants is economically feasible. In Maricopa County, Ariz., near Phoenix, as many as 11 new power plants are in

| Pollutant | Air Pollution Control Technology |
|---|---|
| $NO_x$ | Selective catalytic reduction (SCR)<br>Selective non-catalytic reduction (SNCR)<br>Staged (or multi-staged) combustion<br>Low $NO_x$ burners<br>Flue gas recirculation (FGR)<br>Reduced air combustion |
| VOCs and CO | Regenerative or recuperative thermal oxidation<br>Catalytic oxidation<br>Good combustion practices |
| PM, PM-10, and metals | Baghouse<br>Electrostatic precipitator (ESP)<br>Venturi scrubber |
| $SO_2$ | Packed tower scrubber<br>Fuel switching |

**Table A5-2:** Available control technologies for combustion equipment

| Pollutant | Significant Net Emissions Increase (tons/yr) |
|---|---|
| TSP | 25 |
| PM-10 | 15 |
| $SO_2$ | 40 |
| $NO_2$ | 40 |
| CO | 100 |

**Table A5-3**: PSD significant net emission increases

the development stages. The applicants are rushing to obtain their PSD permits because those that apply later may have an unacceptably small fraction of the PSD increments available for the project as designed and sited. Competition for fractions of PSD increments has influenced power plant planners in other states, including New York and Connecticut.

Because of the time to address these requirements and to accommodate agency and public comment, PSD permits often take one to two years to be issued. Firms should plan to invest the time and effort necessary to obtain a PSD permit or consider appropriate strategies and emission controls in order to avoid its requirements.

New facilities or expansions causing increases in emissions that already exceed NAAQS undergo even greater scrutiny. Such situations are covered by New Source Review (NSR) regulations found at the state level. While similar to PSD in structure, the net proposed increase in annual emissions of the nonattainment pollutant that would draw a facility into NSR is lower than that for triggering PSD. Instead of installing BACT, a facility must install technology to achieve the lowest achievable emission rate (LAER). LAER is the most stringent regulatory standard or demonstrated lowest emission rate regardless of cost or other considerations. In some states, an impact study is required to predict how high the ground-level concentration of the pollutant will rise. Finally, facilities in many states must purchase emission offsets representing the proposed net annual emissions increase of the pollutant plus additional credits to show a net benefit to the environment.

PSD and NSR are of great concern to existing facilities in light of the legal actions EPA initiated against 32 "grandfathered" plants late last year. EPA contends that certain plant owners implemented modifications of combustion equipment, resulting in significant air emission increases, without first obtaining proper permits and review through the PSD and NSR programs. The facilities believe that the modifications were the result of necessary routine maintenance needed for

| NSPS Subpart | Maximum Heat Input | Pollutant | Fuel | Emission Limit (lb/MMBtu) |
|---|---|---|---|---|
| Da | >250 MMBtu/hr | PM | Any | 0.3 and 70% control using liquid fuel |
| | | Opacity | Any | 20% (6 minute average) |
| | | $SO_2$ | Solid | 1.2 and 90% control or 0.6 and 70% control |
| | | $NO_x$ | Solid | 0.50 or 0.60 (based on type of coal) |
| | | | Liquid | 0.50 (shale oil) or 0.30 (others) |
| | | | Gaseous | 0.50 (oil-derived) or 0.20 (others) |
| Db | 100-250 MMBtu/hr | $SO_2$ | Coal/oil | 1.2 (coal)/0.5 (oil) |
| | | PM | Coal | 0.05-0.20 based on co-firing |
| | | Opacity | Any | 20% (6 minute average) |
| | | $NO_x$ | Nat. gas | 0.10 or 0.20 based on release rate |
| | | | Dist. Oil | 0.10 or 0.20 based on release rate |
| | | | Res. Oil | 0.30 or 0.40 based on release rate |
| | | | Coal | 0.5 to 0.8 depending on type |
| Dc | 10-100 MMBtu/hr | $SO_2$ | Coal | 0.5 to 1.2 based on source and mixture |
| | | | Oil | 0.5 |
| | | PM | Coal | 0.05 or 0.10 based on mixture |
| | | Opacity | Any | 20% (6 minute average) |

**Table A5-4:** NSPS requirements for fossil fuel-fired steam generating units

aging equipment and should be exempt from PSD and NSR requirements. Notwithstanding a court's final ruling on the definition of routine maintenance and modification, power generating facilities should review records of historic changes to combustion equipment and assess the impact of the change on air emission rates. Similarly, each future planned change should be investigated for its air emission consequences (Table A5-3).

# NSPS

In the 1970s, EPA began promulgating New Source Performance Standards (NSPS) containing specific emission and monitoring standards for a new installation or a modification of equipment capable of emitting air pollutants from different types of source categories. Contained in 40 CFR Part 60, there are three major NSPS standards that affect the energy community, Subparts Da, Db, and Dc. Emission standards are provided in NSPS for particulate matter (PM), $SO_2$, opacity, and $NO_x$, depending on the fuel combusted and size of unit (Table A5-4). States are instructed not to issue any permits to construct or operate a new or modified source unless it meets NSPS requirements.

# Air permitting

Reviewing emission permits is the principal means of enforcing air pollution regulations. Historically, however, permitting programs have varied widely from state to state. Title V of the 1990 Clean Air Act Amendments developed minimum national air permitting standards. All major sources (facilities) must prepare and submit a Title V permit. Major is defined as having the potential to emit at least 100 tpy of a criteria pollutant (except lead), 10 tpy of any single hazardous air pollutant (a list of 188 compounds deemed hazardous to public health), or 25 tpy of all hazardous air pollutants.

The Title V Operating Permit is meant to be an "umbrella" permit, covering all operations at a major facility. In some states, it will replace all existing individual permits. In others, however, facilities will still be required to prepare and submit individual permit applications for new or modified sources and, if applicable, modify the Title V Operating Permit.

Title V permits undergo stringent agency review and public comment. It is likely that enforcement of air quality rules will be more stringent for facilities with Title V Operating Permits. Most power plants likely exceed at least one major applicability threshold and are subject to the program.

Facilities must meet the terms of the Title V Operating Permit, including all emission and operating limitations. Because it is difficult for a facility to plan for future activities during application preparation, in most states the Title V application contains the opportunity to define Alternate Operating Scenarios (AOS) to anticipate future growth or changes to operation. If emission rates listed in a proposed AOS (due to new equipment, a change in fuel, etc.) comply with all applicable air quality regulations, then a facility can switch to that AOS without pre-approval from the agency. AOSs may be interpreted differently by different states.

Site selection criteria for a prospective new facility should include a state with a less bureaucratic permitting approach.

# Cogen creativity

The goal for a facility permitting a new or modified combustion source is to be allowed to construct or modify and operate the equipment as soon as possible with as few operating constraints as possible for the lowest cost of add-on air pollution control equipment. As an example, the following illustrates a successful strategy to modify a cogeneration facility.

A facility in the Northeastern U.S. had been combusting natural gas and waste gas from its processes in several engine generators for many years to supply electricity and steam. The engines, however, were aging and the plant's electrical demands were growing beyond their capacity, causing the facility to purchase more power from the local utility. The facility decided to purchase new, larger engine generators to meet projected demand 20 years into the future. Since this represented a potential major net increase in emissions of regulated pollutants, the facility, which is in a severe ozone nonattainment region, would have been required to meet LAER restrictions, delaying the project one to two years.

To avoid LAER and speed up the permitting process, the facility agreed to temporary, enforceable usage restrictions on the new engines so that the net emissions increases of the new units would not exceed the applicable PSD or NSR significance limits. The facility still had to purchase energy from the local utility, but a smaller quantity. This underutilization of the new engines was also acceptable because the projected growth in demand had not yet occurred. As a result, the facility received its air permits much faster. The permits contained recordkeeping conditions, such as 12-month rolling average emission limits, to demonstrate that in any given period net emission increases did not rise to levels that would trigger PSD or NSR.

Finally, after the engines were in operation, the facility and its engineer were able to design, install, and operate appropriate air pollution control equipment to enable total energy independence and greater use of the units, while still not exceeding the significance levels. The technologies chosen were less expensive than LAER or BACT. In fact, an additional benefit of this strategy was that while initial permitting was based on conservative vendor guarantees of emission rates, testing under actual conditions showed measured emission rates much lower than the guarantees. This gave the facility more information to design appropriate air pollution control equipment and to re-permit based on more realistic conditions.

# Author —

Marc Karell, P.E., is a Senior Project Engineer at Malcolm Pirnie Inc. in White Plains, N.Y.

# BIBLIOGRAPHY

Altman, R., W. Buckley, and R. Isaac Ray. "Application of Wet Electrostatic Precipitation Technology in the Utility Industry for Multiple Pollutant Control." Power-Gen 2001 International Conference *Proceedings*, Las Vegas, NV, December 11-13, 2001

American Iron and Steel Institute. *Design Guidelines for the Selection of and Use of Stainless Steel.* Distributed by the Nickel Development Institute, publication date unknown

American Iron and Steel Institute. *Welding of Stainless Steel and Other Joining Methods.* Distributed by the Nickel Development Institute, 1979

Amick, P. "Gasification for Power: The Wabash River Project." Paper presented at the First Annual Coal-Gen Conference, Chicago, IL, July 25-27, 2001

Angello, L. "Recent Trends in Gas Turbine Environmental Siting." Power-Gen 2001 International Conference *Proceedings*, Las Vegas, NV, December 11-13, 2001

Austin, F. "Future $SO_2$ and Particulate Control Issues for Coal Plants; Radar Screen vs. Crystal Ball." Power-Gen 2001 International Conference *Proceedings*, Las Vegas, NV, December 11-13, 2001

Barna, K., Walchuk, D., and J. Grusha. "Unique Union of Low $NO_x$ Combustion Controls-Foster Wheeler's TLN System and Duke Power's 'LOFIR' at The Allen and Lee Steam Plants." Power-Gen 2001 International Conference *Proceedings*, Las Vegas, NV, December 11-13, 2001

Begg, E. "What Air Pollution Control Systems Will Your New Coal Plant Require?" Power-Gen 2001 International Conference *Proceedings*, Las Vegas, NV, December 11-13, 2001

Botsford, P.E., C. "The Two Faces of $NO_x$ Control." *Chemical Engineering*, July 2001, pp. 66-71

Bradford, P.E., M., and R. Graver, "Just say $NO_x$.." *Environmental Protection*, December 2000, pp. 31-33, 49

Buecker, B. *Fundamentals of Steam Generation Chemistry.* Tulsa, OK: PennWell Publishing, 2000

Buecker, B. "Gypsum Seed Recycle in Limestone Scrubbers." Paper presented at the Utility Representatives FGD Conference, Farmington, NM, June 10-12, 1986

Buecker, B. *Power Plant Chemistry: A Practical Guide.* Tulsa, OK: PennWell Publishing, 1997

Buecker, B. and J. Meinders. "PRB Coal Switch Not a Complete Panacea." *Power Engineering*, vol. 104, no. 11, November 2000, pp. 76, 78, 80

Buecker, B., J. Wofford, R. DuBose, and D. Ray. "CFB Sorbent Selection Enhances Performance." *Power Engineering*, vol. 101, no. 7, July 1997, pp. 27-29

Cauoette, Todd. Personal conversation. Power Systems Manufacturing, Ltd., Juno Beach, FL

Ciarlante, P.E., V., and M. Zoccola. "Conectiv Energy Successfully Using SNCR for $NO_X$ Control." *Power Engineering*, vol. 105, no. 6, June 2001, pp. 61-62

Cichavowicz, J.E. "What you should know before specifying SCR." *Power*, May/June 1999, pp. 77-82

*COHPAC—Discover the True Benefits of Innovation.* Technical brochure published by Hamon Research-Cottrell

Cooper, J., and D. MacDonald. "Problems and Solutions in Up-grading Air Preheaters for SCR Compatibility." Power-Gen International 2001 Conference *Proceedings*, Las Vegas, NV, December 11-13, 2001

Failing, K.H. and B. Lablanc. "Supercritical Boiler Technology for Clean Coal Power Stations in the USA." Paper presented at the 2001 Power-Gen International Conference, Las Vegas, NV, December 11-13, 2001

Farber, P.E., P., and C. D. Livengood, PhD, "Your Options for Mercury Emissions Control." *Environmental Protection*, August 2000, pp. 26-29

Franz, N. "Cleaning Up the Clean Air Act." *Chemical Engineering*, October 2001, pp. 41-47

Goidich, S. "Technology on the March: Steady Progress in Supercritical Once-Through Technology." Paper presented at Coal-Gen Conference, Chicago, IL, July 25-27, 2001

Karell, P.E., M. "Air Quality Update: Staying Abreast of Changing Regulations." *Power Engineering*, February 2000, pp. 29-32

Kehlhofer, R., R. Bachmann, H. Nielsen, and J. Warner. *Combined-Cycle Gas & Steam Turbine Power Plants*, 2nd ed. Tulsa, OK: PennWell Publishing, 1999

Kokkinos, A., D. Wasyluk, D. Adams, R. Yavorsky, and M. Brower. "B&W's Experience Reducing $NO_x$ Emissions in Tangentially-Fired Boilers-2001 Update." Power-Gen 2001 International Conference *Proceedings*, Las Vegas, NV, December 11-13, 2001

Loper, Ellis. Personal conversation. City Water, Light & Power, Springfield, Illinois

McMillian, R. and D. J. Cramb. "Ultra-Low $NO_x$ Combustor Development for Application to Industrial Gas Turbines." Power-Gen 2001 International Conference *Proceedings*, Las Vegas, NV, December 11-13, 2001

NACE International. **Corrosion Basics: An Introduction**. Houston, TX, 1994

Nicewander, M. *Clean Air Act Permitting: A Guidance Manual*. Tulsa, OK: PennWell Publishing, 1995

Pai, D., and F. Engström. "Fluidised Bed Combustion Technology – The Past, Present, and Future." *Modern Power Systems*, November 1999, pp. 21-26

Powerspan web site, www.powerspan.com

Ross, Jr., R.W. "The Evolution of Stainless Steel and Nickel Alloys in FGD Materials Technology." Paper presented at the EPRI/EPA/DOE 1993 $SO_2$ Control Symposium, Boston, MA, 1993

Schimmoller, B., ed. "Emissions: Compliance Gets Tougher." *Power Engineering*, October 1998, pp. 24-30

Schobert, H. H. *Coal: The Energy Source of the Past and Future*. The American Chemical Society, Washington, D.C., 1987

Sheth, A., and T. Giel. "Understanding the PM-2.5 Problem." *Pollution Engineering*, March 2000, pp. 32-35

Sinder, J.G., ed. *Combustion: Fossil Power*. Windsor, CT: Alstom (formerly Combustion Engineering), 1991

Solt, J.C., and J. Cussen. "Knocking Out $NO_x$." *Environmental Protection*, December 2001, pp. 22-25

Stultz, S. and J.B. Kitto, eds. *Steam*, 40th ed. Barberton, OH: Babcock & Wilcox, a McDermott Company, 1992

Strzelecki, D., Cont. Ed. "On-Site Ammonia-Generating Process Reduces Risks of SCR." *Pollution Engineering*, May 2001, pp. 42-43

Swanekamp, P.E., R., ed. "Project Developers Consider New Solid Fuels, New Technologies." *Power*, March/April 2001, pp. 35-42

Tierney, K., ed. "NAAQS Authority Questioned in Supreme Court." *Pollution Engineering*, October 2000, pp. 9-10

Tran, P., J. Chen, and R. Wolfmensdorf. "Selective Catalytic Reduction of $NO_x$: Commercial applications and a description of the different catalysts used in each application." Power-Gen 2001 International Conference *Proceedings*, Las Vegas, NV, December 11-13, 2001

U. S. Department of Energy. Clean Coal Technology Program: Program Update 2000, Germantown, MD

U. S. Department of Energy. "The Wabash River Coal Gasification Repowering Project." Topical Report, no. 20, September 2000

Usher, M. "Electrostatic Precipitator (ESP) Fundamentals I." Paper presented at the Reinhold Environmental ESP/FF Round Table & Expo 2001, St. Louis, MO, July 22-27

Weilert, C., B. Basel, and P. Dyer. "Startup and Initial Operation of the Wet Limestone FGD Retrofit at V.Y. Dallman Station Units 31 & 32." Power-Gen 2001 International Conference *Proceedings*, Las Vegas, NV, December 11-13, 2001

Weilert, C., D. Randall, and M. Hagan. "Can Existing Air Pollution Control Equipment Meet Future State and Federal Requirements for Mercury Control? - What the ICR Data Tells Us." Power-Gen 2001 International Conference *Proceedings*, Las Vegas, NV, December 11-13, 2001

Wofford, John. Personal conversation, Barlow Projects, Inc.

# INDEX

# A

# P

Particle size data, 34

Particulate control, 41, 113, 118, 134-140, 145: electrostatic precipitation, 134-138; fabric filtration, 134, 138-139

Peat, 60, 63

Performance factors (ESP), 136

Petrochemicals, 67-68

Phosphorous content (steel), 98

Pinch-point temperature, 44-46

Plant efficiency, 42-43

Plaquemine plant (Louisiana), 51

PM2.5 guidelines, 114, 117

Pollutant discharge regulations, 33, 114-117, 143-149

Pollution control, 33, 113-150: air quality and pollutant discharge regulations, 33, 114-117, 143-149; sulfur dioxide control, 118-123; nitrogen oxides control, 123-133; particulate control, 134-139; future issues, 140-141; changing regulations, 143-149

Potassium content, 77, 82

Pour point, 69

Powder River Basin coal switch, 85-89: fuel switch, 85; coal quality and handling, 86-87; boiler issues, 87-88; ash control, 89; other problems, 89

Powerspan system, 141

Precipitator, 134-137, 140

Preheating fuel, 36-37

Pressure vessels, 2

Prevention of significant deterioration (PSD), 143-146, 149

# Q

Quindaro Power Station (Kansas), 85-89

# R

Radiant energy/conduction/convection, 4

Radiant heat, 34-35

Radiant superheater, 11-12

Rank/classification (coal), 61-62

Reagent preparation, 37-38

Reducing atmospheres, 106-107

Refractory, 41

Refuse-derived fuel, 50, 71

Regulations (air quality/pollutant discharge), 33, 114-117, 143-149: changing, 143-149